MACMILLAN/McGRAW-HILL

Math

Macmillan
McGraw-Hill

PROGRAM AUTHORS

Douglas H. Clements, Ph.D.

Professor of Mathematics Education

State University of
New York at Buffalo

Buffalo, New York

Carol E. Malloy, Ph.D.

Assistant Professor of
Mathematics Education

University of North Carolina at
Chapel Hill

Chapel Hill, North Carolina

Lois Gordon Moseley

Mathematics Consultant

Houston, Texas

Yuria Orihuela

District Math Supervisor

Miami-Dade County Public Schools

Miami, Florida

Robyn R. Silbey

Montgomery County Public Schools

Rockville, Maryland

SENIOR CONTENT REVIEWERS

Gunnar Carlsson, Ph.D.

Professor of Mathematics

Stanford University

Stanford, California

Ralph L. Cohen, Ph.D.

Professor of Mathematics

Stanford University

Stanford, California

The McGraw·Hill Companies

Macmillan
McGraw-Hill

Published by Macmillan/McGraw-Hill, of McGraw-Hill Education, a division of The McGraw-Hill Companies, Inc., Two Penn Plaza, New York, New York 10121.

Foldables™, Math Tool Chest™, Math Traveler™, Mathematics Yes!™, Yearly Progress Pro™, and Math Background for Professional Development™ are trademarks of The McGraw-Hill Companies, Inc.

Printed in the United States of America

ISBN 0-02-105014-7/2
   5 6 7 8 9 073 08 07 06

RFB&D 🎧
learning through listening

Students with print disabilities may be eligible to obtain an accessible, audio version of the pupil edition of this textbook. Please call Recording for the Blind & Dyslexic at 1-800-221-4792 for complete information.

Estimate and Measure Length

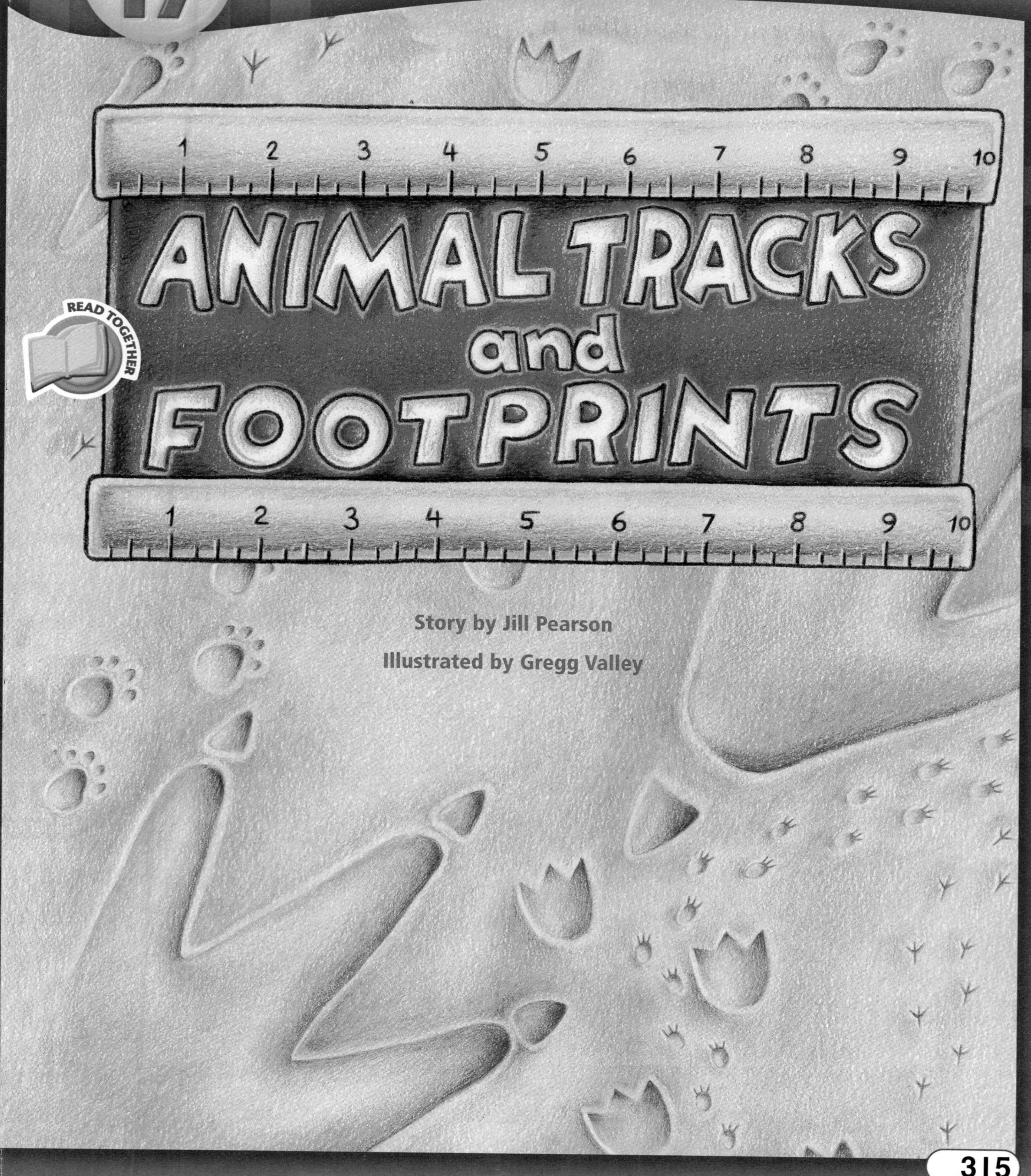

ANIMAL TRACKS
and
FOOTPRINTS

READ TOGETHER

Story by Jill Pearson

Illustrated by Gregg Valley

These are the footprints of a cat.
Is my hand longer or shorter than that?

longer

Next to these bird tracks I put prints of my own feet.

Are the bird's tracks longer or shorter than my footprints? _Shorter_

We find dog tracks on a hill.
We measure them with a shoe from Bill.
Who do you think has the longer foot?

Bill   dog

Name _____

Nonstandard Units of Length

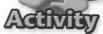
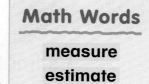

Learn Use ● to **measure** how long the crayon is.

I can estimate first. I think the crayon is about 3 ● long.

Math Words
measure
estimate

Use ● to measure.
Start at the left end.
Put counters side by side. Count how many.

The crayon is about ● long.

Your Turn Estimate. Then use ● to measure.

1

Estimate: about __7__ ●   Measure: about __7__ ●

2

Estimate: about __3__ ●   Measure: about __4__ ●

3

Estimate: about __2__ ●   Measure: about __2__ ●

4 ✏ **Write About It!** How did you use counters to measure a pencil?

Estimate. Then use small ⚬ to measure.

5

Start at the left end to measure.

Estimate: about ___1___ ⚬   Measure: about ___┊___ ⚬

6

Estimate: about ___3___ ⚬   Measure: about ___3___ ⚬

7

Estimate: about ___9___ ⚬   Measure: about ___6___ ⚬

8

Estimate: about ___4___ ⚬   Measure: about ___4___ ⚬

Problem Solving **Critical Thinking**

9 Measure with 🔲.
   Then measure with ⚬.

about _____ 🔲   about _____ ⚬

Are your answers the same or different?

Explain why. _____

THINK
SOLVE
EXPLAIN

Math at Home: Your child used common objects as units to measure length.
Activity: Have your child estimate and then measure the length of a book in pennies.

Name_____

Measure to the Nearest Inch

Learn You can use an inch ruler to measure length. Measure the length of the toy dinosaur.

The dinosaur is about 3 inches long.

Math Words

inch (in.)
length

Your Turn Estimate. Then use an inch ruler to measure.

Object	Estimate	Measure
1	about __4__ inches	about __4__ inches
2	about __2__ inches	about __2__ inches
3	about __3__ inches	about __3__ inches

4. ✏️ Write **About It!** How do you use an inch ruler to measure?

© Macmillan/McGraw-Hill

Find objects like the ones
shown below. Estimate the
length. Then use an inch
ruler to measure.

Object	Estimate	Measure
5 DINOSAURS	about __1__ inches	about __1__ inches
6 (spoon)	about __2__ inches	about __2½__ inches
7 (tube)	about __1½__ inches	about __2__ inches
8 (index card)	about __2__ inches	about __2__ inches

Problem Solving | Critical Thinking

Solve this problem.

9 A large paper clip is 2 inches long.
How long is a chain of 3 large paper clips? __6__ inches

Math at Home: Your child used a ruler to measure inches.
Activity: Have your child show you how to measure the length of a crayon using an inch ruler.

Name_____ **Inch, Foot, and Yard**

HANDS ON
Activity

Learn You can use a yardstick to measure the length of longer objects.

The toy dinosaur is about 2 feet long.

Math Words

foot (ft)

yard (yd)

1 **foot** = 12 inches

1 **yard** = 3 feet

Your Turn Find objects like the ones shown.
Estimate the length or height.
Then use a yardstick to measure.

Object	Estimate	Measure
① 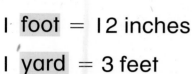	about __22"__	about __23"__
② 	about __8"__	about __8"__
③	about __4"__	about __5"__

④ ✏️ **Write About It!** How many inches are in 3 feet?
How do you know?

© Macmillan/McGraw-Hill

Practice Find an object about as long as the estimate. Then measure.

12 inches = 1 foot
3 feet = 1 yard

Estimate	Name Your Object	Measure
⑤ about 1 inch	_____	about _____
⑥ about 8 inches	_____	about _____
⑦ about 1 foot	_____	about _____
⑧ about 2 feet	_____	about _____
⑨ about 2 feet	_____	about _____

Problem Solving **Mental Math**

⑩ Timmy's shoe is 9 inches long. His brother's shoe is 7 inches long. How much longer is Timmy's shoe?

_____ inches

⑪ A blue rug is 5 feet long. A green rug is 2 feet long. What is the difference between the length of the two rugs?

_____ feet

Math at Home: Your child learned about inches, feet, and yards.
Activity: Have your child show you some items in your home that are about 1 inch, 1 foot, and 1 yard long.

Name_____

Learn You can use a centimeter ruler to measure.
There are 100 centimeters in 1 meter.

The frog is about 8 centimeters long.

Math Words

centimeter (cm)
meter (m)
distance

centimeters
0 1 2 3 4 5 6 7 8

Your Turn Estimate. Use a centimeter ruler to measure.

 1

Estimate: about __10__ cm   Measure: about __10__ cm

 2

Estimate: about __7__ cm   Measure: about __7__ cm

3

Estimate: about __10__ cm   Measure: about __9__ cm

4

Estimate: about __8__ cm   Measure: about __8__ cm

 5 Write **About It!** Would you use centimeters or meters to measure the length of your classroom? Explain.

© Macmillan/McGraw-Hill

Which ruler would you use to measure the length of each object? Draw a line from the object to the ruler.

6

I meter = 100 centimeters.
A meterstick measures 1 meter.

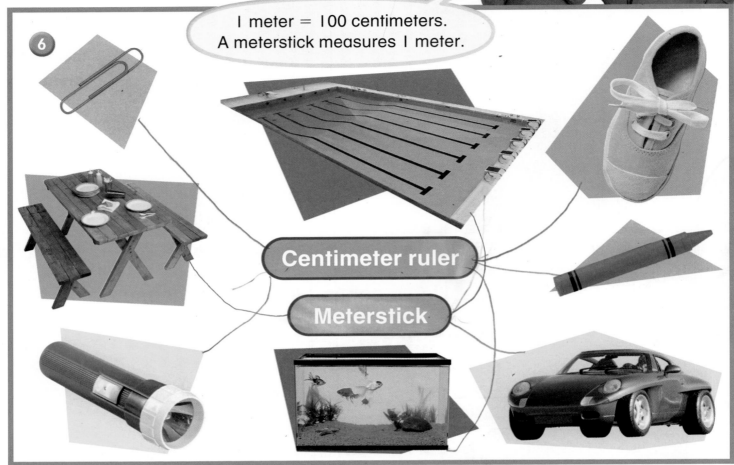

Centimeter ruler

Meterstick

Problem Solving Reasoning

THINK
SOLVE
EXPLAIN

Measure and compare these **distances** in your classroom. Use a meterstick.

7 to

about ___ meters

8 to

about ___ 2 meters

9   Which distance is greater?
Circle the answer that shows the greater distance.

Math at Home: Your child learned about centimeters and meters.
Activity: Have your child show you how to measure a toy car using a centimeter ruler.

324   three hundred twenty-four

Name_____

Dinosaur World

The children see a dinosaur mural.
Three dinosaurs in the mural measure
2 feet long, 25 inches long, and 19
inches long.

Problem Solving

I foot = 12 inches

 **Compare and Contrast**

1 How are the dinosaurs alike? How are they different? Explain.

_____

_____

2 Write the lengths from greatest to least.

_____

3 The mountains in the mural measure 26
inches, 22 inches, 23 inches, and 25 inches
tall. Which are greater than 2 feet tall?

_____

Measuring the Mural

The children want to measure other things in the mural. Three rocks measure 8 inches, 10 inches, and 12 inches long. Then they use a yardstick to measure the length of the whole mural.

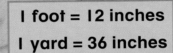

I foot = 12 inches
I yard = 36 inches

 Reading Skill

Compare and Contrast

4 How are the rocks in the mural alike? How are they different? Explain.

_____

_____

_____

5 Which rocks are less than a foot long?

_____

6 The mural measures 46 inches long. How much longer is it than a yard?

_____

 Math at Home: Your child compared and contrasted information to answer questions.
Activity: Have your child measure the lengths of two objects, then compare and contrast the results.

Name_____

Problem Solving Practice

Solve.

1 About how long is this blue mark?
Circle your estimate.

_____

4 inches   3 feet   6 inches   2 meters

2 Estimate how long this toy car is.
Then measure it with a ruler.

Estimate: about _____ inches   Measure: _____ inches

THINK SOLVE EXPLAIN Write **About It!**

3 Measure the dinosaur's footprint.
Use an inch ruler and then a
centimeter ruler. Why are there a
greater number of centimeters
than inches?

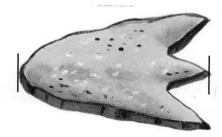

_____

_____

© Macmillan/McGraw-Hill

Problem Solving

Writing for Math

 How long is the dinosaur in the picture?

Think

What can I use to measure?

_____

_____

_____

Solve

I can estimate the length. Then I can measure.

_____

_____

Explain

I can explain how I got the answer.

_____

_____

_____

e-Journal **www.mmhmath.com**
Write about math

Name _____

Estimate the length.
Then use an inch ruler to measure.

Estimate: about _____ inches   Measure: about _____ inches

Estimate: about _____ inches   Measure: about _____ inches

Use a centimeter ruler to measure the length.

Measure: about _____ centimeters

Measure: about _____ centimeters

Solve.

Jay made a paper chain 1 yard long.
Steve made a chain 2 feet long.
Who made the longer chain?
Explain how you know.

Assessment

Test Prep

Choose the best answer.

1 Measure the length of the eraser to the nearest centimeter.

1 centimeter	2 centimeters	3 centimeters	4 centimeters
○	○	○	○

2 27 + 18 = ☐

35	38	45	48
○	○	○	○

3 Write the time.

_____ : _____

4 In which place is the digit 5?

54   _____

THINK SOLVE EXPLAIN

5 Look at this number pattern.
What number would most likely come next?
Explain the pattern.

3, 6, 9, 12, 15, 18

_____

Estimate and Measure Capacity and Weight

SING TOGETHER

Measuring Song

Sung to the tune of "Twinkle, Twinkle, Little Star"

How long is your desk in class?

How much water's in this glass?

How much does an apple weigh?

How cold is it out today?

When you measure, you must choose

What tool is the best to use.

Math at Home

Dear Family,

I will learn about measuring weight, capacity, and temperature in Chapter 18. Here are my math words and an activity that we can do together.

Love, _____

My Math Words

capacity :

the amount a container holds when filled

The capacity is 1 cup.

temperature :

You can use temperature to measure hot or cold.

The temperature is 80°F.

LOG ON
www.mmhmath.com
For Real World Math Activities

Home Activity

Show your child three or four household objects. Ask your child to arrange the objects in order from lightest to heaviest.

© Macmillan/McGraw-Hill

Books to Read

Look for these books at your local library and use them to help your child learn about capacity, weight, and temperature.

- **Counting on Frank** by Rod Clement, Gareth Stevens Publishing, 1991.
- **Lulu's Lemonade** by Barbara deRubertis, The Kane Press, 2000.
- **Room For Ripley** by Stuart J. Murphy, HarperCollins, 1999.

Name_____

Explore Capacity

HANDS ON Activity

Learn You can use a paper cup to measure capacity.

I fill the large cup with dried rice.

This small paper cup holds less rice.

You can fill containers to find the amount they hold.

Math Word

capacity

Your Turn Use a large paper cup to compare. Circle more or less.

1  more (less)

2  more less

3  more less

4  more less

5  more less

6  more less

7 **Write About It!** Choose two other containers. Tell if these hold more or less rice than the large paper cup.

Chapter 18 Lesson 1

three hundred thirty-three **333**

© Macmillan/McGraw-Hill

Practice About how many paper cups does each container hold? You can use pasta or rice to measure.

I can measure to find how many paper cups will fill the bowl.

Container	Estimate	Measure
8		
9		
10		

Problem Solving **Estimation**

THINK
SOLVE
EXPLAIN

11 You can use a chalkboard eraser to measure weight. Hold the eraser in one hand. Hold each object in the other hand. Tell if the object weighs more or less than the eraser.

more (less)

(more) less

Math at Home: Your child learned about capacity.
Activity: Choose three different-sized containers. Have your child use a cup to fill each one with water. Ask which holds the most.

Name Eunice

Fluid Ounce, Cup, Pint, Quart, and Gallon

HANDS ON
Activity

Learn A fluid ounce , cup , pint , quart , and gallon are units of capacity .

	8 fluid ounces = 1 cup	1 cup
	2 cups = 1 pint	1 pint
	2 pints = 1 quart	1 quart
	4 quarts = 1 gallon	1 gallon

Math Words

fluid ounce (fl oz)
cup (c)
pint (pt)
quart (qt)
gallon (gal)
capacity

You can fill containers to find the amount they hold.

Your Turn Complete. You can measure to check.

1. ___8___ fluid ounces = 1 cup

2. ___2___ cups = 1 pint

3. ___2___ pints = 1 quart

4. ___4___ cups = 1 quart

5. ___4___ pints = 2 quarts

6. ___4___ quarts = 1 gallon

7. ✏ Write **About It!** How many cups are in 2 pints? How do you know?

Circle the better estimate.

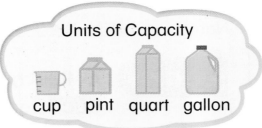
Units of Capacity

cup   pint   quart   gallon

8

more than one quart

(less than one quart)

9

more than 1 cup

(less than 1 cup)

10

(more than one fluid ounce)

less than one fluid ounce

11

more than one gallon

(less than one gallon)

12

more than one pint

(less than one pint)

13

(more than one cup)

less than one cup

x **Algebra • Input/Output Tables**

14 Complete the table.

quarts	1	2	3	4
pints	2	4	6	8
cups	4	8	12	16

Math at Home: Your child learned about fluid ounces, cups, pints, quarts, and gallons.
Activity: Have your child show you some containers in your home that hold 8 fluid ounces, 1 cup, 1 pint, 1 quart, and 1 gallon.

Name _____

Ounce and Pound

HANDS ON Activity

Learn
You can measure weight in ounces and pounds.

The toy bear weighs more than 1 ounce.

The feather weighs less than 1 ounce.

The toothbrush weighs less than 1 pound.

The bowling ball weighs more than 1 pound.

Math Words

ounce (oz)
pound (lb)

16 ounces equal 1 pound.

Your Turn
Circle the better estimate.

1.

more than 1 pound

(less than 1 pound)

2.

more than 1 ounce

less than 1 ounce

3.

more than 1 ounce

less than 1 ounce

4.

more than 1 pound

less than 1 pound

5. Write **About It!** How do you know when an object weighs more than 1 pound?

© Macmillan/McGraw-Hill

Chapter 18 Lesson 3

three hundred thirty-seven **337**

Practice Circle the better unit of measure.

> The notebook weighs I pound.

> The scale is balanced.

6

ounce   (pound)

7

ounce   pound

8

ounce   pound

9

ounce   pound

10

ounce   pound

11

ounce   pound

x Algebra • Missing Numbers

12 How many eggs will balance the scale?

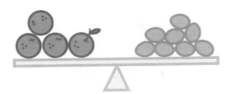

4 🔴 = 8 🔵

2 🔴 = ____ 🔵

Math at Home: Your child determined whether objects weigh more or less than a pound.
Activity: Have your child show you an object in your home that weighs more than a pound and an object that weighs less than a pound.

Milliliter and Liter

Learn A **liter** and a **milliliter** are units of **capacity**.

This bottle holds 1 liter.

A liter is a little more than 1 quart. There are 1,000 milliliters in 1 liter.

Math Words

liter (L)
milliliter (mL)
capacity

This medicine dropper holds 1 milliliter.

Your Turn Circle the better estimate.
You can measure to check.

more than 1 liter
(less than 1 liter)

more than 1 milliliter
less than 1 milliliter

more than 1 liter
less than 1 liter

more than 1 milliliter
less than 1 milliliter

  Write About It! Does a drinking glass hold more or less than 1 liter? Explain.

© Macmillan/McGraw-Hill

Practice Circle the better estimate.
You can measure to check.

There are 1,000 milliliters in 1 liter.

6

(about 2 liters)

about 20 liters

7

about 5 milliliters

about 50 milliliters

8

about 2 liters

about 20 liters

9

about 2 liters

about 20 liters

x Algebra • Missing Addends

Draw a picture to solve.
Write a number sentence.

10 Donna filled 5 glasses with 1 liter of juice. How many glasses could she fill with 3 liters of juice?

_____ + _____ + _____ = _____

Show Your Work

Math at Home: Your child learned about milliliters and liters.
Activity: Have your child find one container that holds more than 1 liter and one container that holds less than 1 liter.

Name_____

Gram and Kilogram

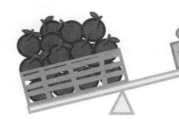

HANDS ON Activity

Learn You can measure mass in grams and kilograms.
There are 1,000 grams in 1 kilogram.

Math Words

gram (g)
kilogram (kg)

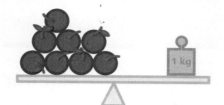

**less than
1 kilogram**

**about the same
as 1 kilogram**

**more than
1 kilogram**

Your Turn Circle the better estimate.

1

lighter than 1 kilogram
(heavier than 1 kilogram)

2

(lighter than 1 kilogram)
heavier than 1 kilogram

3

(lighter than 1 kilogram)
heavier than 1 kilogram

4

lighter than 1 kilogram
(heavier than 1 kilogram)

 5 Write **About It!** Is a large object always heavier than
a small object? Explain.

Circle the better estimate.
Then use a balance scale
to measure.

An apple is less
than I kilogram.

Object	Estimate	Measure
6	lighter than I kilogram (heavier than I kilogram)	lighter than I kilogram (heavier than I kilogram)
7	lighter than I kilogram (heavier than I kilogram)	lighter than I kilogram heavier than I kilogram
8	(lighter than I kilogram) heavier than I kilogram	lighter than I kilogram heavier than I kilogram
9	(lighter than I kilogram) heavier than I kilogram	lighter than I kilogram heavier than I kilogram

Problem Solving **Critical Thinking**

10 Which is heavier, I kilogram of rocks or
I kilogram of feathers? Explain.

Nethar because there both one
Kilogram.

Math at Home: Your child learned about grams and kilograms.
Activity: Have your child name one object that is less than a kilogram and one object that is more than a kilogram.

Name _____

Learn Temperature can be measured in degrees Fahrenheit (°F).

Math Words

temperature
degrees Fahrenheit (°F)
degrees Celsius (°C)

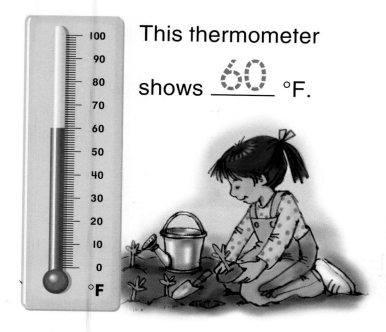

This thermometer shows __60__ °F.

This thermometer shows __82__ °F.

Try It Write each temperature.

1

__55__ °F

2

__18__ °F

3 Write **About It!** What do you think the temperature is today?
Look at the thermometer to check.

© Macmillan/McGraw-Hill

Practice Write each temperature.

Temperature can be measured in degrees Celsius (°C), too.

4
_18_ °C

5
_32_ °C

6
_16_ °C

7
_2_ °C

Problem Solving **Critical Thinking**

Show Your Work

THINK
SOLVE
EXPLAIN

8 Read the thermometer. Draw a picture. Show what you would wear to go outside. Write about your picture.

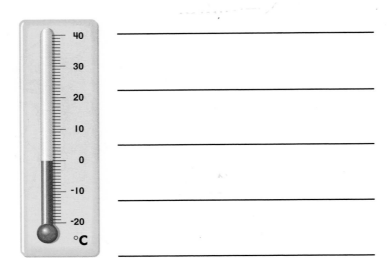

_____

_____

_____

_____

_____

Math at Home: Your child learned about temperature.
Activity: Ask your child what you should wear when the temperature is 25 °F.

344 three hundred forty-four

Name_____

Problem Solving Strategy

Use Logical Reasoning

You can use logical reasoning to help you solve problems. Carrie wants to know the temperature outside. She needs to decide which tool to use.

cup

balance scale

thermometer

Problem Solving

Read

What do I already know?   _____

What do I want to find out?   _____

What do I need to find out?   _____

Plan

I can decide which tool measures temperature.

Solve

I can carry out my plan.

A _____ measures temperature.

Look Back

Does my answer make sense?   Yes   No

Why? _____

Circle the correct tool to measure.

Problem Solving

1 How heavy is it?

2 How cold is it?

3 How much water does it hold?

4 How heavy is it?

Math at Home: Your child learned about measurement tools.
Activity: Ask your child what tools he or she would use to measure different items in your home.

Game Zone

Practice at School ★ Practice at Home

How Big?

▶ Each partner chooses a ruler.
▶ Find one object to match each length.
▶ Write what you found.
▶ The first player to find 3 matching lengths wins.

 2 players

You Will Need

inch ruler

centimeter ruler

Length	Found Object about this Size
Player 1	
about 2 inches	
about 7 inches	
about 4 inches	
Player 2	
about 2 cm	
about 7 cm	
about 18 cm	

© Macmillan/McGraw-Hill

Technology Link

Addition and Subtraction • Calculator

Use a to find the new temperature.

The temperature is 62°F.
It goes down 9°F.
What is the new temperature?

Press.

You see 53.

The new temperature is 53°F.

---

1. The temperature is 95°F.
   It goes down 6°F.
   What is the new temperature?

   Press. _____ _____ _____

   _____°F is the new temperature.

---

2. The temperature is 58°F.
   It goes up 6°F.
   What is the new temperature?

   Press. _____ _____ _____

   _____°F is the new temperature.

3. Write About It! Why would you use a calculator or mental math to solve these problems?

Name_____

Circle the better estimate.

1

more than I ounce
less than I ounce

2

more than I cup
less than I cup

3

more than I pound
less than I pound

4

more than I kilogram
less than I kilogram

5

about 2 liters
about 200 liters

6

about 2 milliliters
about 200 milliliters

Write each temperature.

7    _____ °F

    8    _____ °C

9 Circle the tool you would use
to measure how heavy a shoe is.

thermometer   cup   scale   centimeter ruler

10 Danny fills a jug with 4 pints of water. Does
he have more or less than I quart of water?
Explain your answer.

_____

Assessment

Spiral Review and Test Prep
Chapters 1—18

Choose the best answer.

1 There are 72 oranges in the box. The clerk takes out 27 oranges to make juice. How many oranges are left?

45 oranges   55 oranges   85 oranges   99 oranges
   ⬭   ⬭   ⬭   ⬭

2 Jan has 49¢. Which group of coins could she have?

8 nickels, 4 pennies

1 quarter, 2 dimes, 4 pennies

1 quarter, 2 dimes, 9 pennies

2 quarters

   ⬭   ⬭   ⬭   ⬭

3 A cat weighs about _____.

5 ounces   10 ounces   9 pounds   100 pounds
   ⬭   ⬭   ⬭   ⬭

4 Name an item that is about 1 meter long. _____

5 Write the number that makes the number sentence true.

$56 = \underline{\quad} + 10$

6 Sue started her homework at 3:30. She finished at 5:30. How long did she spend doing homework? Explain your answer.

_____

_____

Name_____

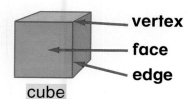 These are 3-dimensional figures.
Some have a **face**, an **edge**, and a **vertex**.

Math Words

face	sphere
edge	cylinder
vertex	cone
cube	pyramid
rectangular prism	

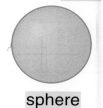

— **vertex**
— **face**
— **edge**

cube

A cube has 6 faces,
12 edges, and 8 vertices.

  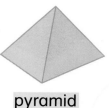 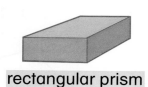

sphere   cylinder   cone   pyramid   rectangular prism

Try It   Circle the objects that have
the same 3-dimensional figure.

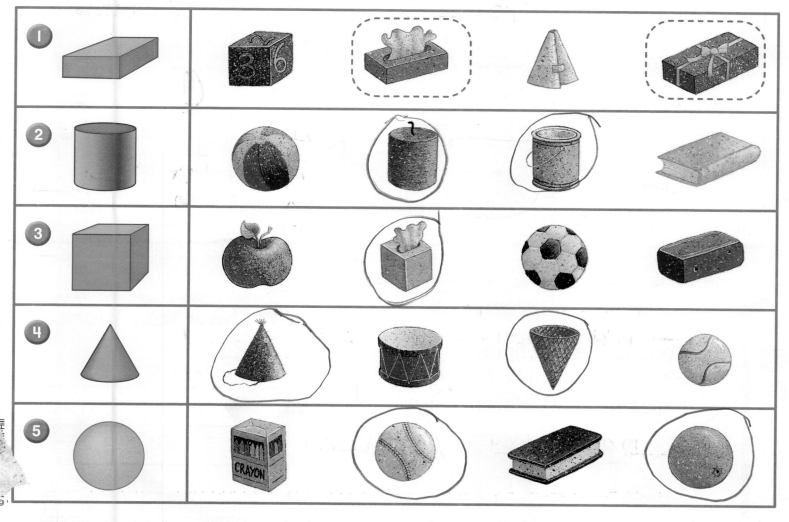

6 ✎ Write **About It!** Describe the 3-dimensional figure made
when you put two cubes together.

Practice

Circle the 3-dimensional figure named. Write how many faces, vertices, and edges it has.

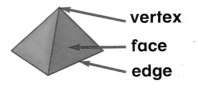

3-Dimensional Figures		Faces	Vertices	Edges
7 rectangular prism		6	8	12
8 cone		1	1	3
9 cylinder		2	0	2
10 cube		6	8	12

Problem Solving · Visual Thinking

Show Your Work

11 Look at the two figures.
How are they alike?
How are they different?

_____

_____

_____

_____

Name_____

2-Dimensional Shapes

 **Learn** These are 2-dimensional shapes.
Most have sides and angles.

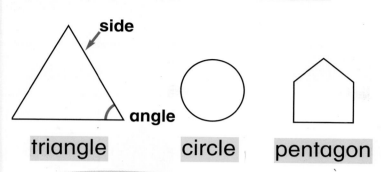

triangle   circle   pentagon

These shapes are quadrilaterals. Each has 4 sides and 4 angles.

square   rectangle   parallelogram

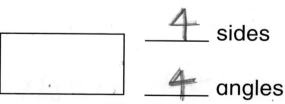

Try It Tell how many sides and angles.
Name each 2-dimensional shape.

1. 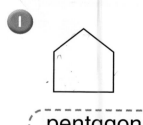 _5_ sides
_5_ angles
(pentagon) square

2. _4_ sides
_4_ angles
triangle   rectangle

3. _4_ sides
_4_ angles
parallelogram   rectangle

4. _3_ sides
_3_ angles
triangle   quadrilateral

5. Write **About It!** How are squares, rectangles, and parallelograms alike?

Practice

Color the shape named .
Then tell how many sides and angles each has.

	Color the shape.	Sides	Angles
6 pentagon		5	5
7 parallelogram		___	___
8 triangle		___	___
9 square		___	___

Problem Solving Visual Thinking

THINK SOLVE EXPLAIN

10 This shape is an octagon.
Tell two things about it.

_____

_____

_____

_____

Math at Home: Your child learned about triangles, quadrilaterals, and pentagons.
Activity: Have your child find or draw examples of a triangle, a rectangle, and a square.

Name _____

HANDS ON
Activity

Learn Find a 2-dimensional shape in a 3-dimensional figure. You can trace a face.

All the faces of the cube are squares.

Your Turn Find objects like these 3-dimensional figures. Trace around the bottom face. Circle the 2-dimensional shape you made.

 1
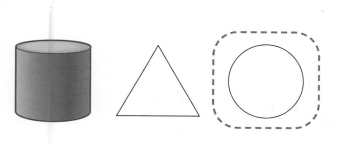

 2
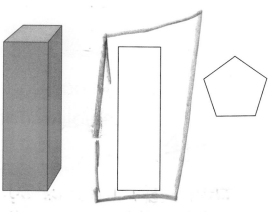

3

4

 5 Write **About It!** You traced around the face of a 3-dimensional figure to make a circle. What 3-dimensional figure did you trace?

Practice Circle the 3-dimensional figures you can use to make all the 2-dimensional shapes.

2-Dimensional Figures	3-Dimensional Figures
6	
7	
8	
9	

Problem Solving **Visual Thinking**

THINK
SOLVE
EXPLAIN

10 Compare the ball and the circle.
How are they alike and different?

They are both round
The ball is 3-dimenchinel and the circle is 2-dimenchin

11 Compare the block and the square.
How are they alike and different?

They are both square
The block is 3-dimenchinel and the square is 2-dimenchinel

Math at Home: Your child learned about the shapes of faces of 3-dimensional objects.
Activity: Have your child trace the faces of a box and a can and name the 2-dimensional shapes he or she traced.

Name_____ **Combine Shapes**

Learn You can put two shapes together to make a new shape.

A trapezoid has 4 sides. A hexagon has 6 sides.

trapezoid   hexagon

Math Words

trapezoid
hexagon

I can use 2 trapezoids to make a hexagon.

Your Turn Use pattern blocks to make new shapes. Then complete the chart.

Use these pattern blocks.	Draw a new shape.	How many sides?	How many angles?	Name of new shape.
1	(hexagon outline)	6	6	hexagon
2	(quadrilateral)	4	4	dima rombus

3 ✏️ Write **About It!** Can you use squares to make a rectangle? Explain.

© Macmillan/McGraw-Hill

Practice Use pattern blocks to make new shapes. Complete the chart.

Use these pattern blocks.	Make a new shape.	How many sides?	How many angles?	Name of new shape.
4 ◻◻	⬚	4	4	rectangle
5 △ ◆	(trapezoid drawing)	4	4	trapezoid
6 ⬢ ▲▼	(trapezoid drawing)	4	4	hexagon

Problem Solving **Visual Thinking**

Show Your Work

7 Tommy made this shape using 3 pattern blocks. What were the blocks he used? Draw lines to show how Tommy put the blocks together.

_____

Math at Home: Your child combined shapes to make new shapes.
Activity: Have your child show you how to use 2 triangles to make a new shape.

360 three hundred sixty

Name_____

Add or subtract.
Circle the ones that
need regrouping.

46
− 23

23
+ 17
40

60
− 12

74
− 18

26
− 19

49
+ 11

36
+ 36

18
+ 21

16
+ 34

40
− 15

©Macmillan/McGraw-Hill

Draw the hour and minute hands to show the time.

1

2

3

4

5

6

7

8

9

10

Math at Home: Your child practiced telling time.
Activity: Ask your child to tell you when it is 2:50. Repeat with different times.

Name_____

Shape Patterns

Learn You can make a pattern of shapes.
Use pattern blocks. Then use letters
to show the pattern another way.

Math Word

unit

unit

(A B) A B A B A B

To find a pattern, look
for the shapes that repeat. The
shapes that repeat make a unit.

Your Turn Use pattern blocks to show the pattern.
Then circle the pattern unit.

1

2

3

4

5

6 ✏ Write **About It!** How can you tell which shapes come next?

© Macmillan/McGraw-Hill

Practice Use pattern blocks to show the pattern. Then use letters to show each pattern another way.

7
A B A B A B A B

8
A A B A A B A A B

9
A B C A B C A B C

10
A B A A B A A B A

Make it Right

Draw a picture to solve.

11 What is wrong with Meg's pattern? Circle the shape that is wrong.

Draw the correct pattern below.

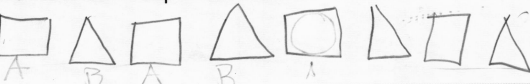

A B A B A

 Math at Home: Your child learned about shape patterns.
Activity: Have your child draw a shape pattern, using squares and triangles.

Name_____

The City

Min and Anna build a model of their city. They use 3-dimensional figures for the buildings and trees. They use 2-dimensional shapes for signs.

 Reading Skill

Use Illustrations

Use the picture to help answer the questions.

1. Name two 3-dimensional figures that were used in the city. How are they alike? How are they different?

   _____

   _____

   _____

2. Which 2-dimensional shapes were used for the signs in the city?

   _____

3. How are the signs the same? How are they different?

   _____

   _____

The Museum and the Park

Anna made a museum and park in the city. How many different kinds of shapes did she use?

Use Illustrations

Use the picture to help answer the questions.

4 Name the 2-dimensional shapes that were used in the park.

_____

_____

5 Find the 2-dimensional sign. What shape is it?

_____

6 Choose 2 of the 2-dimensional shapes you named. Tell how they are the same and how they are different.

_____

_____

_____

Math at Home: Your child used illustrations to help him or her answer questions.
Activity: Ask your child to draw some shapes, then use the pictures to explain how the shapes are alike and different.

Name_____

Solve.

1 Kathy painted this picture.

How many ☐ ? _____

How many △ ? _____

2 Chris and Beth painted this picture of a bus. Look for the 2-dimensional shapes in the picture.

How many squares? _____

How many circles? _____

THINK SOLVE EXPLAIN **Write a Story!**

3 Write a story about the shapes in this picture.

_____

_____

_____

_____

_____

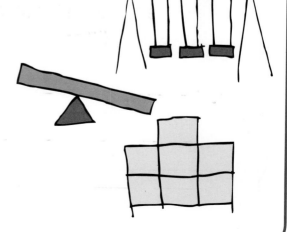

© Macmillan/McGraw-Hill

Writing for Math

Write about the object in the picture.
Use the words on the cards to help.

faces	vertices	edges

Writing

Think

I can tell the shape of the object.

What do I know about 3-dimensional figures?

_____

_____

Solve

I can look at the object and name the figure.

_____

Explain

I can tell you how I know the figure.

_____

_____

_____

Journal **www.mmhmath.com**
Write about math

Name_____ **Congruence**

Learn Congruent shapes have the same size and shape.

These squares are congruent. They have the same size and shape.

congruent

not congruent

Math Word

congruent

Your Turn Use pattern blocks to find the congruent shapes.

Use this pattern block.	Color the two congruent shapes.
1	
2	
3	

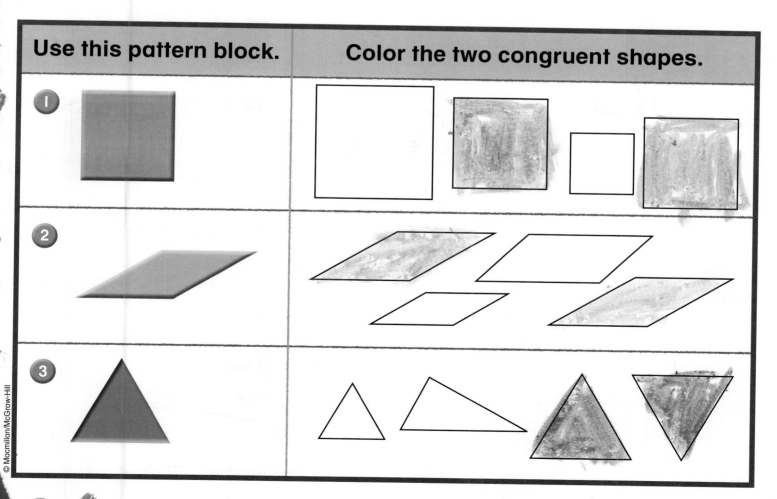

4 ✎ Write **About It!** When are two 2-dimensional shapes congruent?

Practice Find the two congruent shapes. Then color them.

These hexagons are the same size and shape. They are congruent.

5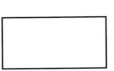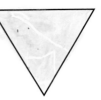

6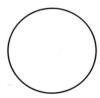

7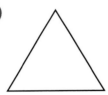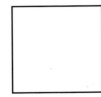

8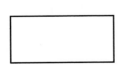

Spiral Review and Test Prep

Choose the best answer.

9 Which shape has 5 sides and 5 angles?

◯ ◯ ◯

10 Which number completes the addition?

$\boxed{} + 30 = 60$

70   50   40   30
◯   ◯   ◯   ◯

 Math at Home: Your child learned about congruent shapes.
Activity: Have your child show you two congruent shapes.

374 three hundred seventy-four

Name _____ **Symmetry**

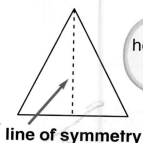

Learn Some shapes have a line of symmetry.

These shapes have a line of symmetry. The two parts match exactly.

Math Word

line of symmetry

These shapes do not have a line of symmetry.

line of symmetry

Your Turn Make each paper shape. Fold each shape to make a line of symmetry. Draw the line in the fold.

1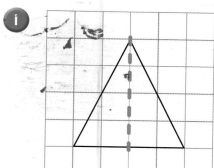

2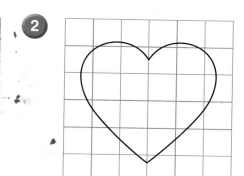

3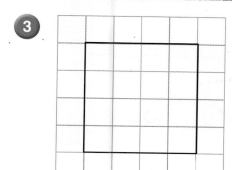

4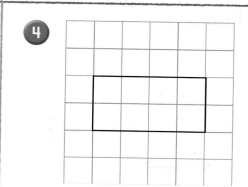

5  **Write About It!** What can you tell about a shape that has a line of symmetry?

When 2 parts match exactly there is a line of symmetry.

6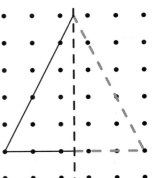

7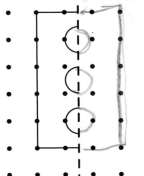

8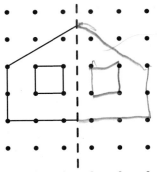

9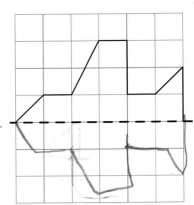

10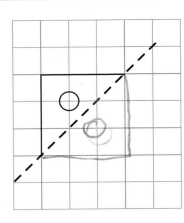

11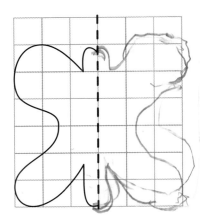

Problem Solving **Visual Thinking**

12 Draw a picture to show the reflection.

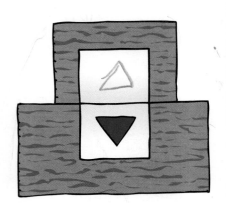

Math at Home: Your child learned about lines of symmetry.
Activity: Show your child a paper plate, a fork, and a knife. Ask him or her if these shapes have lines of symmetry.

Name_____

Slides, Flips, and Turns

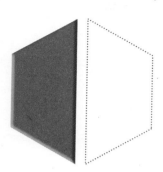 
HANDS ON
Activity

Learn You can move shapes in different ways.

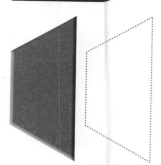   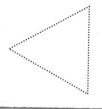 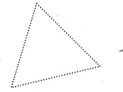 

Math Words
slide
flip
turn

This is a **slide**.   This is a **flip**.   This is a **turn**.

This is a flip.

Your Turn Color to show the move.
Use the pattern blocks.

1. Which shows a slide?

  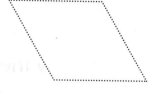 

2. Which shows a flip?

  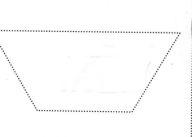 

3. Which shows a turn?

4. ✎ Write **About It!** How are a flip, slide, and turn alike?

© Macmillan/McGraw-Hill

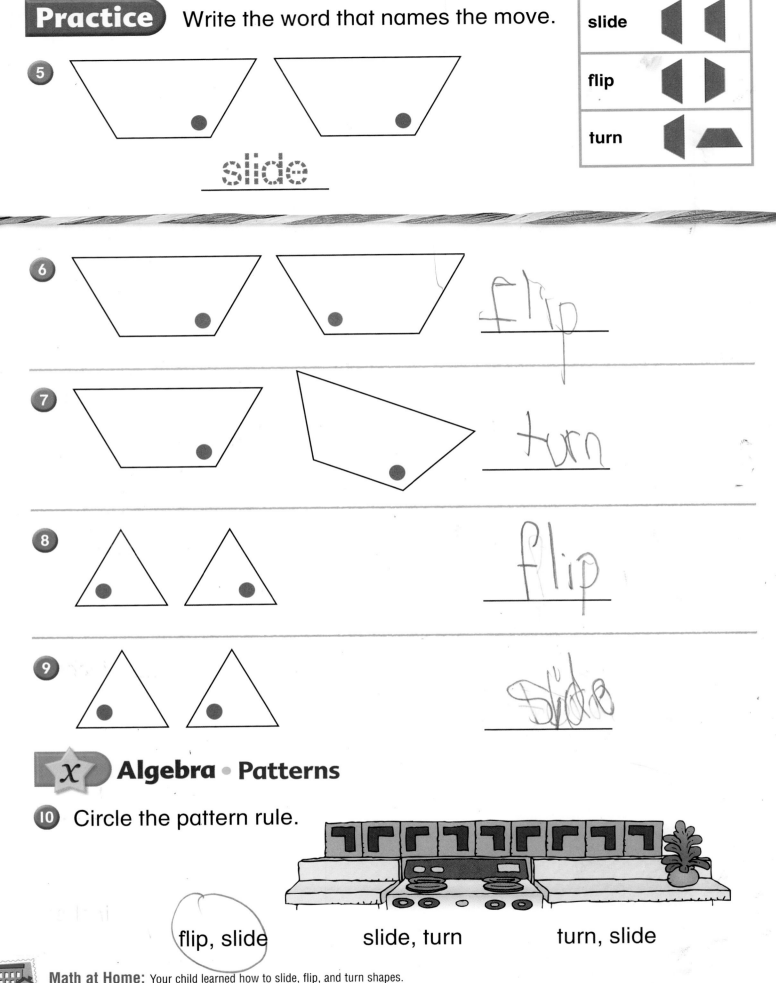

Practice Write the word that names the move.

slide	
flip	
turn	

5. _slide_

6. flip

7. turn

8. flip

9. slide

⚡ **Algebra • Patterns**

10. Circle the pattern rule.

(flip, slide)   slide, turn   turn, slide

Math at Home: Your child learned how to slide, flip, and turn shapes.
Activity: Cut out a paper triangle and have your child slide, flip, and turn this shape.

378 three hundred seventy-eight

Name_____ **Perimeter**

 Learn The distance around a shape is its perimeter.
Measure each side. Use an inch ruler.

Math Word

perimeter

2 inches

1 inch **1 inch**

2 inches

I can measure
the perimeter.

__2__ + __1__ + __2__ + __1__ = __6__

The perimeter is __6__ inches.

 **Your Turn** Find objects like the one shown.
Use an inch ruler to find the perimeter.

_____ + _____ + _____ + _____ = _____ inches

_____ + _____ + _____ + _____ = _____ inches

_____ + _____ + _____ + _____ = _____ inches

  Write About It! How do you find the perimeter of a sheet
of paper?

Practice Find each perimeter. Use an inch ruler.

The distance around a shape is its perimeter.

5 What is the perimeter of the Reptile House?

_____ + _____ + _____ + _____ + _____ = _____ inches

6 What is the perimeter of the sign?

_____ + _____ + _____ + _____ = _____ inches

Problem Solving (Critical Thinking

7 Draw 2 different shapes that have the same perimeter.

Math at Home: Your child found the perimeter of shapes.
Activity: Have your child find the perimeter of a book in your home.

380 three hundred eighty

Name_____ **Area**

Learn Area is the number of square units
that covers a shape.

Math Word
_____
area

The area of the
green rectangle is

 square units.

Your Turn Find the area of each shape. Use pattern
block squares to cover each shape.

1.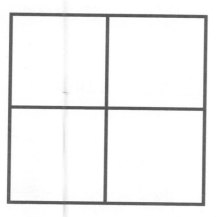

_____ square units

2.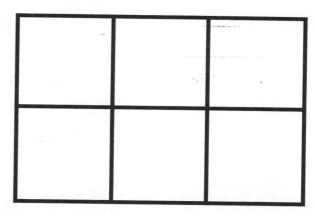

_____ square units

3.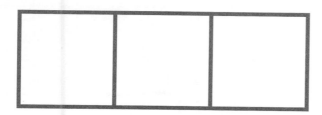

_____ square units

4. ✏️ **Write About It!** How do you find the area of a shape?

Practice Find the area of each shape.
Use pattern block squares.

Area is measured in
square units.

5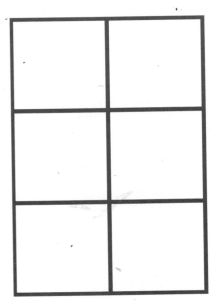

___6___ square units

6

___4___ square units

7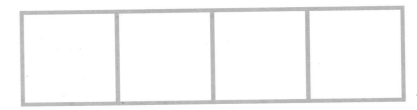

___4___ square units

Problem Solving **Visual Thinking**

8 About how many square units are
inside the large triangle? Explain.

_____ square units

_____

_____

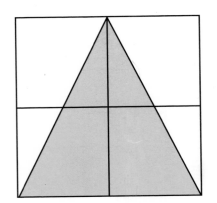

Math at Home: Your child learned about area.
Activity: Have your child find the area of a rectangular pan using square units, such as square crackers.

Problem Solving

Read ⟩ Plan ⟩ Solve ⟩ Look Back

Problem Solving Strategy

Name_____

Guess and Check

You can guess and check to help you solve problems.

Casey drew a square garden. The perimeter is 20 centimeters. How long is each side?

Read

What do I already know? _20_ centimeters is the perimeter.

What do I need to find out? _The length of the side of the garden._

Plan

I know a square has 4 equal sides. I can guess and check to find the length of each side. *because* *I know*

Solve

I can carry out my plan. I guess the length of the side. Then I add the lengths to check.

First try:
2 cm + 2 cm + 2 cm + 2 cm = 8 cm. No.

Second try:
5 cm + 5 cm + 5 cm + 5 cm = 20 cm. Yes.

Look Back

Each side is 5 centimeters long.
Does my answer make sense?   Yes   No

How do I know? _____

© Macmillan/McGraw-Hill

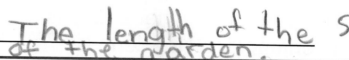

Problem Solving

Guess and check to solve.

Draw or write to explain.

1. José draws a square house. The perimeter of the house is 12 centimeters. How long is each side of the house?

   _____ centimeters

   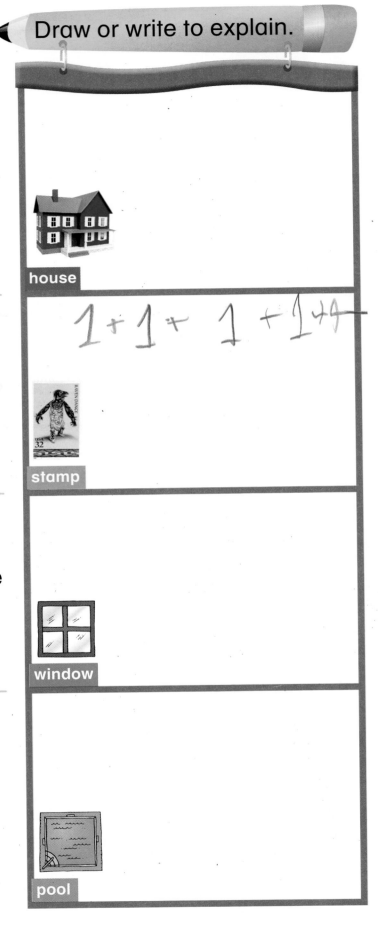

   house

   $1 + 1 + 1 + 1 + 4$

2. Ashley draws a square stamp. The perimeter of her stamp is 4 centimeters. How long is each side of the stamp?

   __1__ centimeter

   stamp

3. Tyler draws a square window. The perimeter of the window is 20 inches. How long is each side of the window?

   _____ inches

   window

4. Alexis draws a square pool. The perimeter of the pool is 16 inches. How long is each side of the pool?

   _____ inches

   pool

Math at Home: Your child solved problems by using the guess-and-check strategy.
Activity: Have your child tell you how he or she solved exercise 4.

Name_____

Build It

▶ Take turns. Each player covers part of the picture with one of the pattern blocks.

▶ Continue until the picture is completed.

▶ The winner is the player who places the last pattern block on the picture.

 2 players

You Will Need

pattern blocks

Technology Link

Shapes • Computer

Use 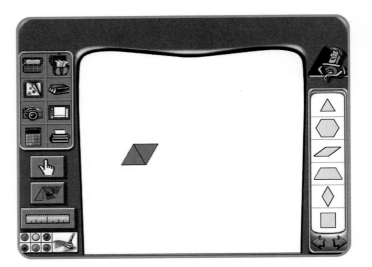 to make shapes.

- Choose a mat to show shapes.

- Stamp out 2 △.

- Turn I △ 2 times.

- Put the △ △ together.

  You made a parallelogram.

Make other shapes and put them together.
You can use the computer or pattern blocks.

1 Stamp out 2 ▱.

Flip I ▱.

Slide I ▱.

Show what you made.

2 Stamp out 3 ▱.

Turn I ▱.

Flip I ▱.

Show what you made.

 For more practice use Math Traveler.™

Name _____

At this museum, you can see two giant spheres.
What shapes do you see in the picture?
Draw them in the space below.

Fold down

Museum Math

READ TOGETHER

Museums are fun places to visit.
You can learn many things from them.

In many museums, you can visit old Native American homes.

What shapes do you see?

This outdoor museum has butterflies! The butterflies show symmetry.

One museum has the machine that took astronauts to the moon. They measured what they found there.

Name ______________________________

Linking Math and Science

Weight and Soils

The Earth is covered with different kinds of soil.

Science Words

topsoil
sandy soil

Topsoil

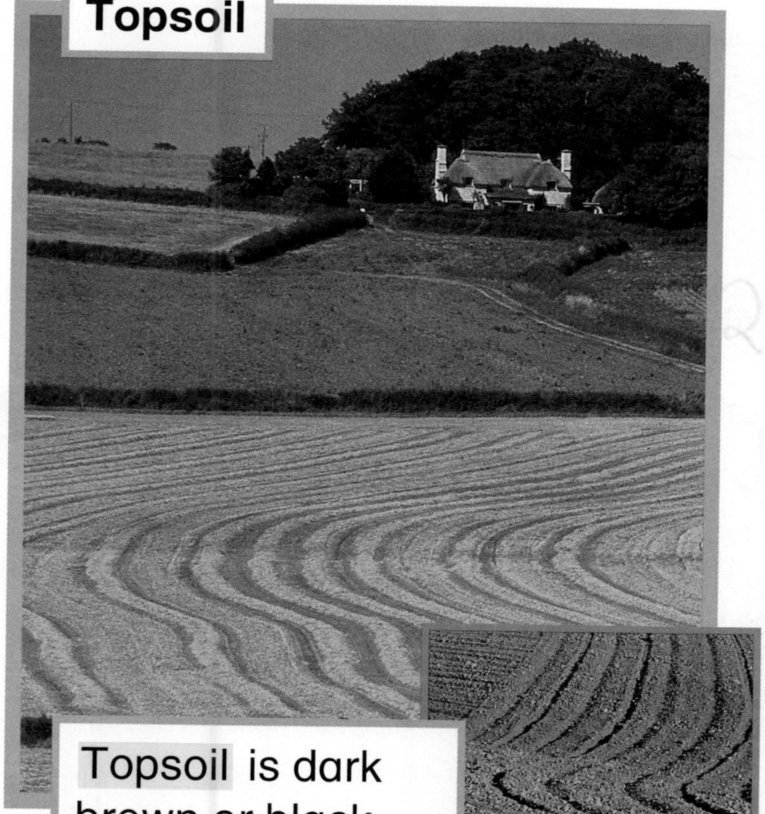

Topsoil is dark brown or black. It can hold water.

Sandy Soil

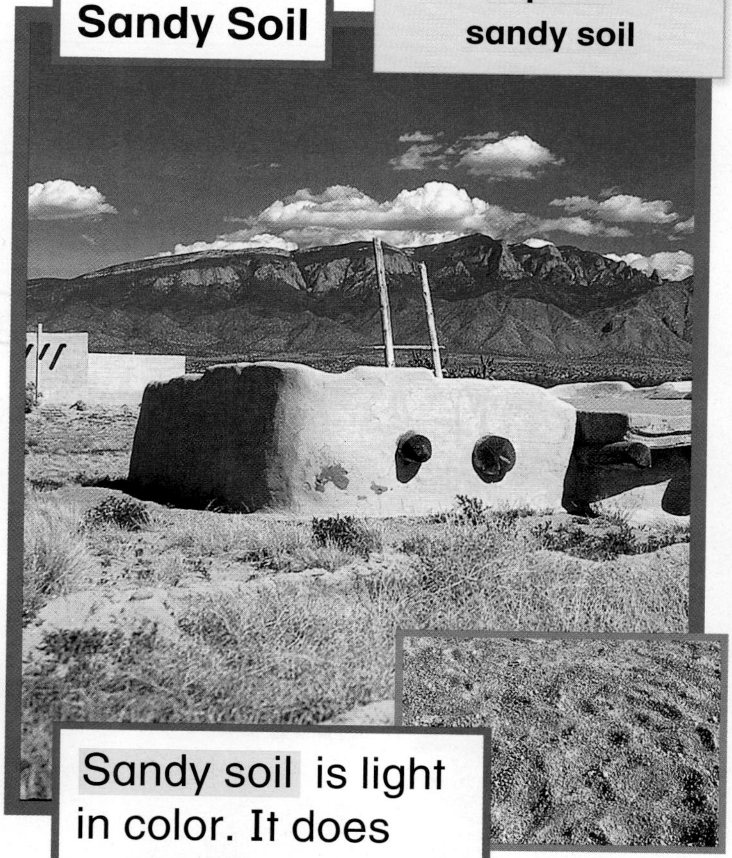

Sandy soil is light in color. It does not hold water well.

Problem Solving

Use the pictures.
Circle the word to complete each sentence.

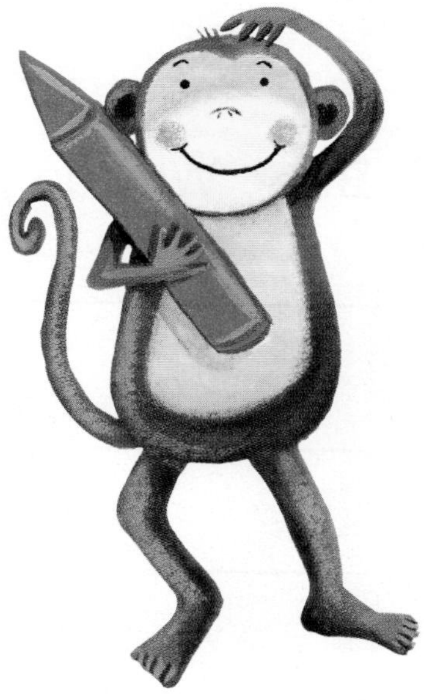

1. A soil that doesn't hold water well is ____.

   topsoil   sandy soil

2. A soil that can hold water is ____.

   topsoil   sandy soil

What to Do

You Will Need

sandy soil
topsoil
2 same-size
cups
balance

- **Observe** Tell how each soil feels.

_____

_____

_____

- Fill a cup with sandy soil. Label it.

- Fill the other cup with topsoil. Label it.

- **Measure** Put each cup on the balance.

Problem Solving

Solve.

 Compare Which soil is heavier?

 Infer Why do you think that soil is heavier?

_____

_____

_____

Math at Home: Your child applied measurement concepts to investigate properties of soil.
Activity: Repeat this activity with your child using different kinds of soils from around your home.

390 three hundred ninety

Name

Math Words

Draw lines to match.

1 4 cups

2 48° F

3 12 inches

| I foot |
| I quart |
| temperature |

Skills and Applications

Measurement (pages 317–322)

Examples

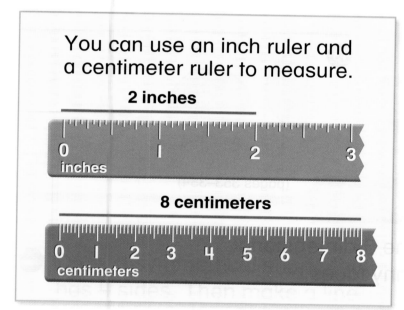

You can use an inch ruler and a centimeter ruler to measure.

2 inches

8 centimeters

4 _____ centimeters

5 _____ centimeters

6 _____ inches

You can decide the better unit of measure.

 (ounce) pound

 gram (kilogram)

7  ounce pound

8 pen gram kilogram

9 feather ounce pound

Unit Review

Unit 5
Enrichment

What is the Volume?

You Will Need

cubes

Volume is the amount of space an object takes up. You can find the volume of a figure.

First Estimate how many cubes are in this figure.

Estimate: about ___20___ cubes

Next Build the figure. Write how many cubes you used.

Count: There are ___18___ cubes.

Estimate the number of cubes. Then build the figure. Write how many cubes you used.

 1

Estimate: about _____ cubes

Count: There are _____ cubes.

 2

Estimate: about _____ cubes

Count: There are _____ cubes.

3

Estimate: about _____ cubes

Count: There are _____ cubes.

 4

Estimate: about _____ cubes

Count: There are _____ cubes.

UNIT 6

CHAPTER 21

Place Value to Thousands

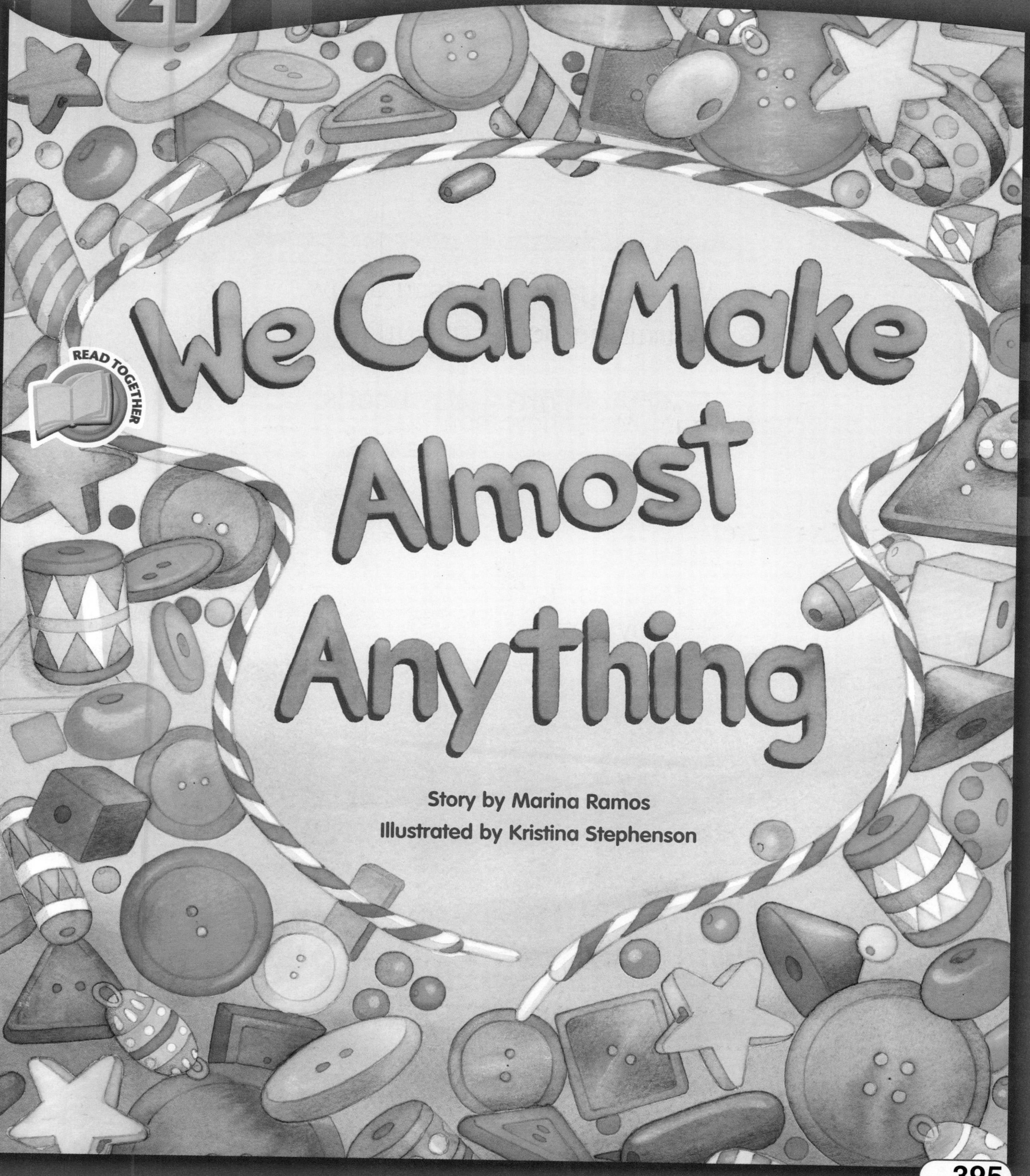

READ TOGETHER

We Can Make Almost Anything

Story by Marina Ramos

Illustrated by Kristina Stephenson

We string the beads in a row.
Count the beads as you go.

Now there are _____ beads.

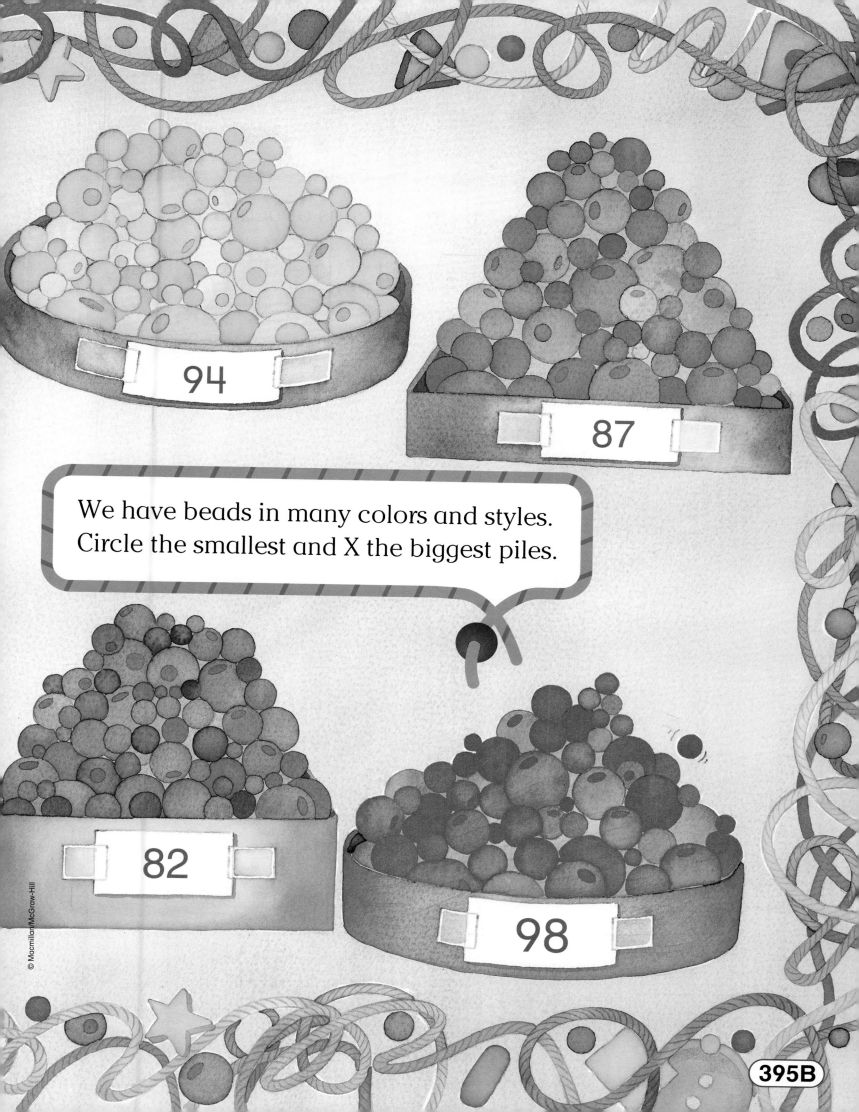

We have beads in many colors and styles. Circle the smallest and X the biggest piles.

94

87

82

98

If you make these 2 strings of beads,

how many beads will you need? _____ beads

395C

We have piles of beads, big and small.
Circle the pile with the most beads in all.

Math at Home

Dear Family,

I will learn about place value to thousands in Chapter 21. Here are my math words and an activity that we can do together.

Love, _____

My Math Words

digits :
used when representing a number

124

the digits are 1, 2, 4.

place value :
the amount that each digit in a number stands for

1 → hundreds   2 → tens   4 → ones

Home Activity

Put about 100 beans, beads, or buttons in a bowl.

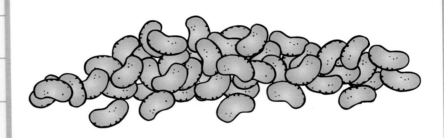

Have your child find different ways to group and count the beans, such as by 5s or 10s.

© Macmillan/McGraw-Hill

Books to Read

Look for these books at your local library and use them to help your child learn place value to thousands.

- **Only One** by Marc Harshman, Cobblehill Books, 1993.
- **The 329th Friend** by Marjorie Weinman Sharmat, Four Winds Press, 1992.
- **Moira's Birthday** by Robert N. Munsch, Firefly Books Ltd., 1988.

LOG ON
www.mmhmath.com
For Real World Math Activities

Name_____ **Hundreds**

Learn You can group tens to make hundreds.

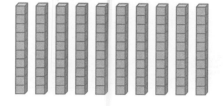 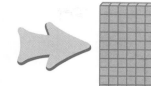

Math Word

tens
hundreds

___10___ tens   =   ___1___ hundred = __100__ in all

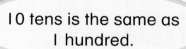

10 tens is the same as 1 hundred.

Your Turn Use models to make groups of hundreds. Write how many hundreds.

1. 20 tens = __2__ hundreds

2. 30 tens = _____ hundreds

3. 40 tens = _____ hundreds

4. 50 tens = _____ hundreds

5. 60 tens = _____ hundreds

6. 70 tens = _____ hundreds

7. 80 tens = _____ hundreds

8. 90 tens = _____ hundreds

9. Write **About It!** How many hundreds are there in 900? How do you know?

Write how many tens and how many ones. Then write the number.

You can trade
1 hundred for 10 tens.

10

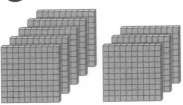

8 hundreds = __80__ tens = __800__ ones = __800__

11

4 hundreds = _____ tens = _____ ones = _____

12

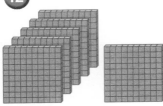

6 hundreds = _____ tens = _____ ones = _____

13

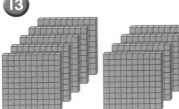

9 hundreds = _____ tens = _____ ones = _____

Problem Solving | **Number Sense**

14 Randy trades 10 hundreds for 1 thousand.

10 hundreds is 1,000.

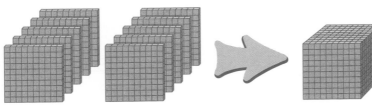

Write the number. _____

Math at Home: Your child used models to make hundreds.
Activity: Ask your child to count by hundreds to 1,000.

Name_____

Hundreds, Tens, and Ones

Learn You can use hundreds, tens, and ones to show 325.

  5

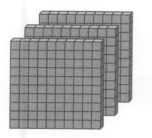

There are 3 hundreds, 2 tens, and 5 ones.

hundreds	tens	ones
3	2	5

 hundreds tens 5 ones

Try It Write how many hundreds, tens, and ones.

1. 458

____4____ hundreds ____5____ tens ____8____ ones

hundreds	tens	ones
4	5	8

2. 142

_____ hundred _____ tens _____ ones

hundreds	tens	ones

3. 636

_____ hundreds _____ tens _____ ones

hundreds	tens	ones

4. ✏ Write **About It!** How many hundreds, tens, and ones are in 572? Explain how you know.

Write how many hundreds, tens, and ones.

Remember to write a zero when there are no tens or ones.

5 630

_____6_____ hundreds _____3_____ tens _____0_____ ones

hundreds	tens	ones
6	3	0

6 246

_____ hundreds _____ tens _____ ones

hundreds	tens	ones

7 515

_____ hundreds _____ ten _____ ones

hundreds	tens	ones

Write each number.

8 3 hundreds 2 tens 6 ones

_____

9 5 hundreds 0 tens 8 ones

_____

Problem Solving — **Number Sense**

10 Look at these two numbers.

What does the 0 stand for in each number?

305 350

_____

Math at Home: Your child learned about hundreds, tens, and ones in 3-digit numbers.
Activity: Write a 3-digit number such as 291. Ask your child to tell how many hundreds, tens, and ones.

Name_____

Learn You can learn the place value of a digit by its place in a number.

Math Words

place value

digit

expanded form

hundreds	tens	ones
▦	|||||	▪ ▪ ▪ ▪ ▪
2	5	7

2 hundreds 5 tens 7 ones

200 + 50 + 7

257

You can write a number in expanded form.

Try It Write how many hundreds, tens, and ones. Then write the number.

1. 3 hundreds 1 ten 5 ones

hundreds	tens	ones
3	1	5

300 + 10 + 5

315

2. 3 hundreds 2 tens 2 ones

hundreds	tens	ones

____ + ____ + ____

____

3. 2 hundreds 4 tens 8 ones

hundreds	tens	ones

____ + ____ + ____

____

4. 1 hundred 0 tens 7 ones

hundreds	tens	ones

____ + ____ + ____

____

5. ✏️ **Write About It!** How can you tell the value of a digit in a 3-digit number?

© Macmillan/McGraw-Hill

 Write how many hundreds, tens, and ones. Then write the number.

The value of the digit 6 in 672 is 600.

6 6 hundreds 7 tens 2 ones

hundreds	tens	ones
6	7	2

600 + 70 + 2

672

7 8 hundreds 3 tens 1 one

hundreds	tens	ones

___ + ___ + ___

_____

8 9 hundreds 2 tens 5 ones

hundreds	tens	ones

___ + ___ + ___

_____

9 7 hundreds 4 tens 0 ones

hundreds	tens	ones

___ + ___ + ___

_____

Circle the value of the green digit.

10 591

(500) 50 5

11 256

600 60 6

12 924

200 20 2

13 485

800 80 8

14 372

200 20 2

15 273

200 20 2

Math at Home: Your child learned the place value of 3-digit numbers.
Activity: Say a 3-digit number and have your child tell you the value of each digit.

Name_____

Compare. Use < or >.

Color > .

Color < .

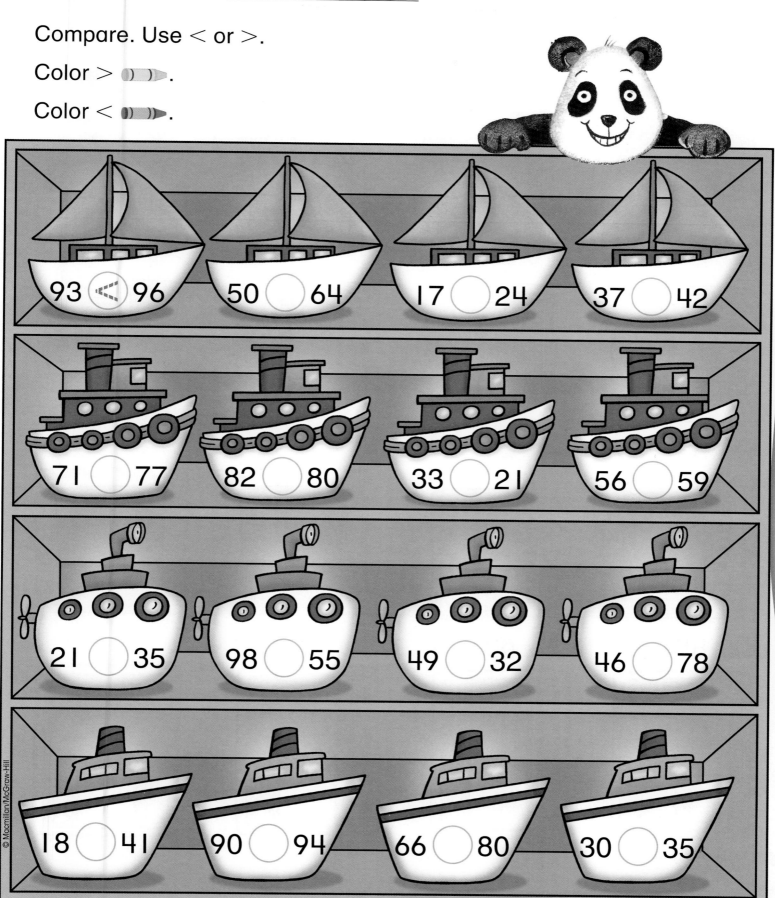

93 < 96   50 ◯ 64   17 ◯ 24   37 ◯ 42

71 ◯ 77   82 ◯ 80   33 ◯ 21   56 ◯ 59

21 ◯ 35   98 ◯ 55   49 ◯ 32   46 ◯ 78

18 ◯ 41   90 ◯ 94   66 ◯ 80   30 ◯ 35

© Macmillan/McGraw-Hill

Extra Practice

Write the numbers in order from least to greatest.

1 44   22   57   40

2 77   92   58   31

3 33   39   11   24

4 85   81   99   63

 www.mmhmath.com
For more Practice

Math at Home: Your child practiced comparing and ordering 2-digit numbers.
Activity: Ask your child to arrange the numbers in exercise 2 on this page from greatest to least.

404   four hundred four

Name_____

Learn You can use models and a place-value chart to explore numbers to thousands.

Math Word

thousands

1 thousand 2 hundreds 4 tens 5 ones

thousands	hundreds	tens	ones
1	2	4	5

You can show numbers in different ways.

__1,000__ + __200__ + __40__ + __5__   __1,245__

Word name: one thousand two hundred forty-five

Try It Write how many thousands, hundreds, tens, and ones. Then write the number.

① 2 thousands 1 hundred   6 tens   8 ones

thousands	hundreds	tens	ones
2	1	6	8

__2,000__ + __100__ + __60__ + __8__   __2,168__

② 1 thousand 1 hundred   3 tens   4 ones

thousands	hundreds	tens	ones

_____ + _____ + _____ + _____   _____

③ ✎ Write **About It!** Tell what you know about the number 2,561.

© Macmillan/McGraw-Hill

Write a zero when there are no hundreds, tens, or ones.

4 2 thousands 1 hundred   0 tens   5 ones

thousands	hundreds	tens	ones

_____ + _____ + _____ + _____   _____

5 1 thousand 3 hundreds   8 tens   2 ones

thousands	hundreds	tens	ones

_____ + _____ + _____ + _____   _____

Use the number line. Write the missing number in the pattern.

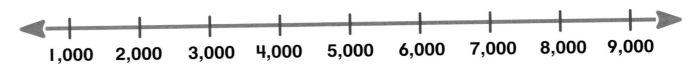

1,000   2,000   3,000   4,000   5,000   6,000   7,000   8,000   9,000

6 1,000, 2,000, 3,000, _____, 5,000, 6,000, 7,000, _____ , 9,000

Problem Solving **Number Sense**

THINK
SOLVE
EXPLAIN

7 Pat has 982 beads. Does she have more or less than 1,000 beads?

Explain. _____

_____

Math at Home: Your child explored numbers to the thousands.
Activity: Write a 4-digit number. Have your child tell you what he or she knows about the number.

406 four hundred six

Name_____

Dana's Marbles

Dana collects things. She has marbles in 4 different colors. She keeps each color in a separate jar. Dana has 98 red, 136 purple, 87 yellow, and 69 pink marbles.

 Reading Skill **Find the Main Idea**

1. What is the story about?

   _____

2. Write the numbers of marbles in order from greatest to least.

   _____

3. Write how many hundreds, tens, and ones for the number of purple marbles.

   _____ hundred _____ tens _____ ones

4. If someone gave Dana 10 more yellow marbles, how many yellow marbles would she have in all?

   _____

Problem Solving

Writing for Math

Write a story about the number of beads in the picture.

Use the words on the cards to help.

| hundreds | tens | ones |

Think

How many jars of 100 beads do I see in the picture? _____ jars

How many bracelets of 10 beads do I see in the picture?

_____ bracelets

How many other beads do I see in the picture? _____

_____ hundreds _____ tens _____ ones = _____

Solve

I can write my story now.

_____

_____

_____

Explain

I can tell you how my number matches my story.

_____

_____

e-Journal **www.mmhmath.com**
Write about math

Chapter 21 Writing for Math

Name_____

Write how many tens and how many ones.
Then write the number.

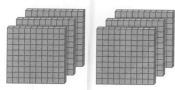

6 hundreds = _____ tens = _____ ones = _____

5 hundreds = _____ tens = _____ ones = _____

Write how many hundreds, tens, and ones.
Then write the number.

3   2 hundreds 8 tens 2 ones

hundreds	tens	ones

_____ + _____ + _____

_____

4   4 hundreds 0 tens 8 ones

hundreds	tens	ones

_____ + _____ + _____

_____

5   Jack has 200 blue marbles, 315 gold marbles, and 18 pink marbles. If Terry gives him 100 more blue marbles, how many blue marbles will he have in all?

_____ blue marbles

Math at Home

Dear Family,

I will learn about comparing and ordering 3-digit numbers in Chapter 22. Here are my math words and an activity that we can do together.

Love, _____

My Math Words

is greater than > :

27 > 26

is less than < :

26 < 27

is equal to = :

26 = 26

Home Activity

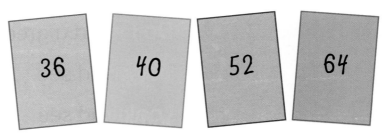

Write a few 2-digit numbers that are not in order. Have your child write the numbers from least to greatest.

| 36 | 40 | 52 | 64 |

Repeat the activity with other sets of numbers.

© Macmillan/McGraw-Hill

Books to Read

Look for these books at your local library and use them to help your child learn number relationships and patterns.

- **Benjamin's 365 Birthdays** by Judi Barrett and Ron Barrett, Atheneum Books, 1974.

- **How Much, How Many, How Far, How Heavy, How Long, How Tall Is 1000?** by Helen Nolan, Kids Can Press, 1995.

- **What's Next, Nina?** by Sue Kassirer, The Kane Press, 2001.

www.mmhmath.com
For Real World Math Activities

Name_____

Compare Numbers

Learn You can compare numbers using >, <, or =.

Math Words

is greater than >
is less than <
is equal to =

Compare 324 and 133.

First compare the hundreds.

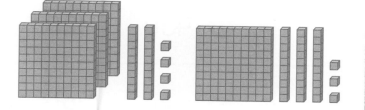

324 is greater than **133**.

324 > 133

Compare 213 and 231.

The hundreds are the same. Compare the tens.

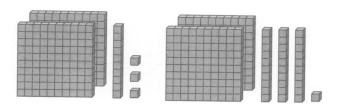

213 is less than **231**.

213 < 231

Try It Compare. Write >, <, or =.

① 347 > 197	② 415 ◯ 429	③ 621 ◯ 635
④ 320 ◯ 352	⑤ 850 ◯ 750	⑥ 583 ◯ 540
⑦ 958 ◯ 779	⑧ 207 ◯ 210	⑨ 295 ◯ 232
⑩ 853 ◯ 853	⑪ 923 ◯ 927	⑫ 567 ◯ 567

⑬ **Write About It!** Explain how you would compare 157 and 162.

Practice

Write >, <, or =.
Compare 128 and 124.

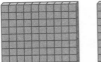   

128 is greater than 124.

128 > 124

> The hundreds are the same. The tens are the same. Compare the ones.

⑭ 689 (>) 627

⑮ 372 ◯ 374

⑯ 450 ◯ 425

⑰ 281 ◯ 182

⑱ 105 ◯ 105

⑲ 789 ◯ 799

⑳ 601 ◯ 601

㉑ 233 ◯ 230

㉒ 955 ◯ 955

㉓ 723 ◯ 723

㉔ 325 ◯ 300

㉕ 252 ◯ 251

㉖ 533 ◯ 515

㉗ 142 ◯ 180

㉘ 697 ◯ 655

Problem Solving · Logical Reasoning

㉙ I am greater than 3 hundreds 2 tens and 2 ones. I am less than 3 hundreds 2 tens and 4 ones. What number am I?

_____

㉚ I am greater than 8 hundreds 7 tens and 5 ones. I am less than 8 hundreds 7 tens and 7 ones. What number am I?

_____

 Math at Home: Your child learned how to compare 3-digit numbers.
Activity: Ask your child to name two numbers that are greater than 286 and then two numbers that are less than 550.

Name_____

Learn You can use a number line to put numbers in order.

419 is just before 420.

421 is just after 420.

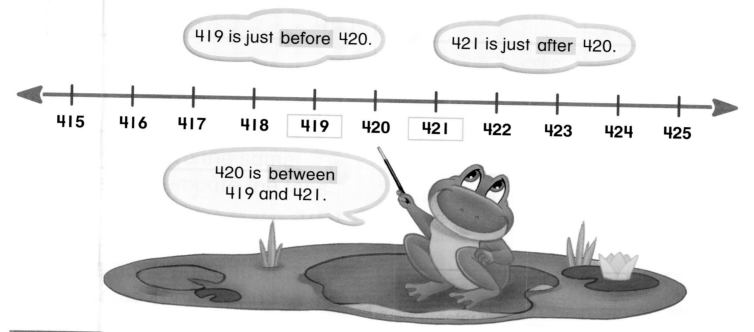

415 416 417 418 419 420 421 422 423 424 425

420 is between 419 and 421.

Try It Write the number that is just before.

1
118 119 120

2
_____ 723 724

Write the number that is just after.

3
258 259 _____

4
538 539 _____

Write the number that is between.

5
345 _____ 347

6
598 _____ 600

7 Write **About It!** How would you find the number that is just before 900?

Chapter 22 Lesson 2

four hundred seventeen **417**

Practice Order the numbers from least to greatest.

6 | 172 236 242 221 | _172_ , _221_ , _236_ , _242_

7 | 327 518 354 569 | _____ , _____ , _____ , _____

8 | 436 431 620 400 | _____ , _____ , _____ , _____

9 | 721 514 726 743 | _____ , _____ , _____ , _____

Order the numbers from greatest to least.

10 | 523 357 525 341 | _525_ , _523_ , _357_ , _341_

11 | 619 630 647 632 | _____ , _____ , _____ , _____

12 | 871 287 284 835 | _____ , _____ , _____ , _____

13 | 901 918 908 981 | _____ , _____ , _____ , _____

 Make it Right

 THINK SOLVE EXPLAIN

14 Petro put numbers in order from greatest to least.

Why is Petro wrong? Make it right.

528, 438, 440

Name_____   **Number Patterns**

Learn   You can use number patterns to
help you count.

Count by hundreds.
Each number is
100 more.

| 150 | 250 | 350 | 450 | 550 | 650 | 750 |

Count by tens.
Each number is
10 more.

| 113 | 123 | 133 | 143 | 153 | 163 | 173 |

Count by ones.
Each number is
1 more.

| 321 | 322 | 323 | 324 | 325 | 326 | 327 |

Try It   Write the missing number.
Then circle the counting pattern.

Numbers	Pattern: Count by		
① 140, 150, 160, _170_, 180	hundreds	(tens)	ones
② 365, 465, _____, 665, 765	hundreds	tens	ones
③ _____, 235, 236, 237, 238	hundreds	tens	ones
④ 528, 538, 548, _____, 568	hundreds	tens	ones

⑤ Write **About It!** How can you tell if a number pattern is
counting by hundreds?

Practice

Write the missing number.
Then circle the counting pattern.

Count by tens.

428, 438, 448, 458, 468

Numbers	Pattern: Count by		
6 920, 930, __940__, 950, 960	hundreds	(tens)	ones
7 432, 532, 632, _____, 832	hundreds	tens	ones
8 785, _____, 787, 788, 789	hundreds	tens	ones
9 142, 143, _____, 145, 146	hundreds	tens	ones
10 500, 510, 520, _____, 540	hundreds	tens	ones
11 299, 399, 499, 599, _____	hundreds	tens	ones

Problem Solving Critical Thinking

12 The art store orders 10 brushes each month. How many brushes does the shop order in 5 months?

_____ brushes

Number of Months	Brushes
1	10
2	20
3	30
4	40
5	?

Math at Home: Your child described and extended number patterns.
Activity: Pick a number between 100 and 500. Have your child count by ones, tens, or hundreds.

Name_____

Learn You can use a number line to count forward or backward.

Count forward by tens.

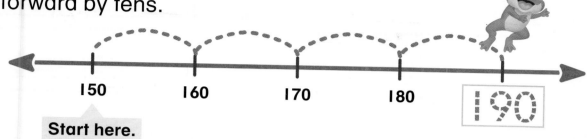

150 160 170 180 190

Start here.

Count backward by tens.

150 160 170 180 190

Start here.

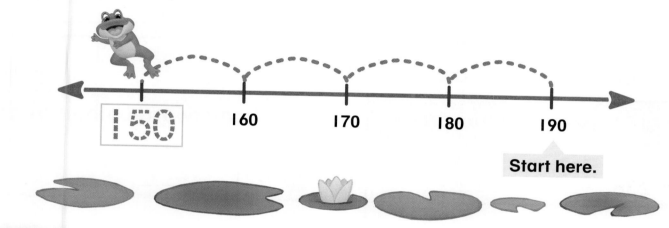

Try It Count forward or backward.

Count forward by tens.	Count backward by tens.
① 256, __266__, __276__	⑤ 371, __361__, __351__
② 827, _____, _____	⑥ 594, _____, _____
③ 479, _____, _____	⑦ 643, _____, _____
④ 132, _____, _____	⑧ 238, _____, _____

⑨ **Write About It!** What number would you get if you count forward by ten from 299? Explain.

© Macmillan/McGraw-Hill

Practice   Count forward or backward by ones.

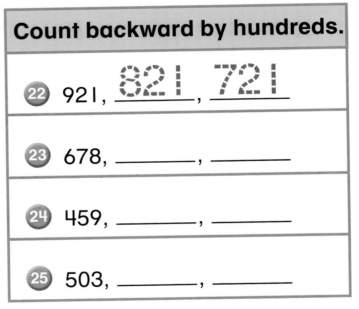

Count forward by ones.

10 347, __348__, __349__

11 862, _____, _____

12 648, _____, _____

13 599, _____, _____

Count backward by ones.

14 105, __104__, __103__

15 424, _____, _____

16 200, _____, _____

17 701, _____, _____

Count forward or backward by hundreds.

Count forward by hundreds.

18 238, __338__, __438__

19 417, _____, _____

20 672, _____, _____

21 800, _____, _____

Count backward by hundreds.

22 921, __821__, __721__

23 678, _____, _____

24 459, _____, _____

25 503, _____, _____

Problem Solving   Number Sense

26 Look at the number.
    Complete the charts.

357

10 more	_____
10 less	_____

100 more	_____
100 less	_____

Math at Home: Your child learned to count forward and backward.
Activity: Pick a 3-digit number. Ask your child to count forward and backward by tens. Continue with other numbers.

Problem Solving Strategy

Name_____

Make a Table • Algebra

Sometimes you can make a table to help solve a problem.

Briana and Amber want to make 5 bracelets. They need 10 beads for each bracelet. How many beads do they need in all?

Read

What do I already know? _____ bracelets

_____ beads for each bracelet

What do I need to find? _____

Plan

I can make a table. Then look for the pattern in the table.

Solve

I can carry out my plan.

The girls need ___50___ beads.

Number of Bracelets	1	2	3	4	5
Number of Beads	10	20			

Look Back

How can I check my answer?_____

_____

Problem Solving

Complete the table to solve.

Problem Solving

1. Jessica puts a dime in her piggy bank every week. How much can she save in 5 weeks?

_____ ¢

Weeks	1	2	3	4	5
Money	10¢				

2. David collects toy cars. He can fit 6 toy cars on a shelf. How many toy cars can he fit on 5 shelves?

_____ toy cars

Shelves	1	2	3	4	5
Cars					

3. Joe's class uses 1 box of paper every week. There are 100 sheets of paper in a box. How many sheets of paper do they use in 5 weeks?

_____ sheets

Weeks	1	2	3	4	5
Sheets					

4. 50 mugs are sold at the craft fair every day. How many mugs are sold in 5 days?

_____ mugs

Days	1	2	3	4	5
Mugs					

Math at Home: Your child learned to solve problems by making a table.
Activity: Make a chart for a problem such as, "Jan's family makes a road trip. They drive 100 miles a day. How far do they travel in 5 days?" Help your child fill in the chart to solve.

Name_____

Game Zone

Practice at School ★ Practice at Home

Trading Tens

▶ Choose a workmat. Take turns.

▶ Toss the 🎲. Put that many tens on your workmat.

▶ When you have 10 ▭, trade them for 1 ▪.

▶ The first player to get 5 ▪ wins.

2 players

You Will Need

🎲

20 ▭

9 ▪

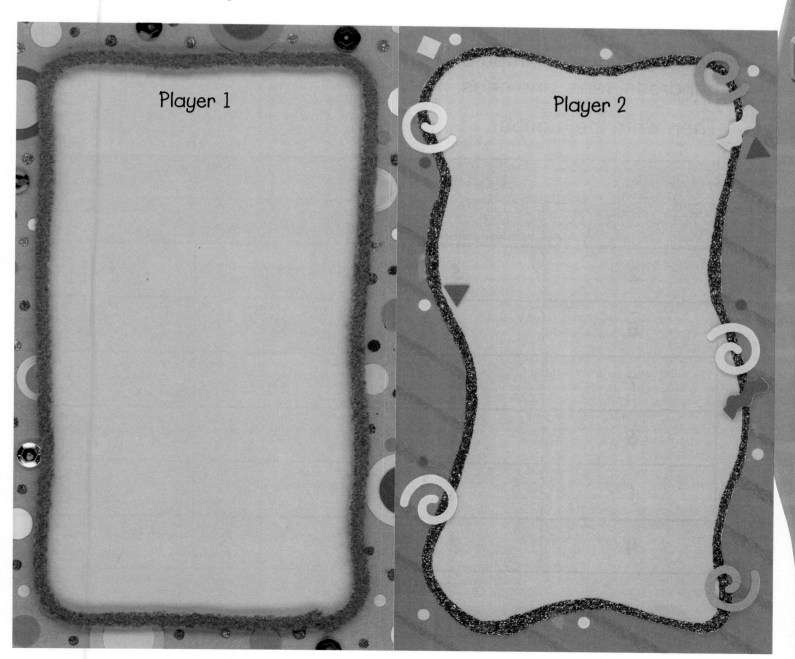

Player 1

Player 2

© Macmillan/McGraw-Hill

Technology Link

Place Value • Computer

Use to make numbers.

Choose a mat to show place value.

Stamp out 3 ▦.

Stamp out 5 ▭.

Stamp out 9 ◻.

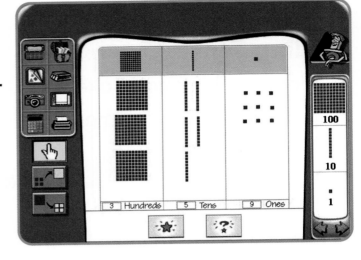

What is the number? __359__

You can use the computer to stamp out hundreds, tens, and ones.

Then write the number.

Stamp Out			What is the Number?
▦	▭	◻	
2	6	1	261
5	0	2	
7	8	3	
6	2	0	
1	5	8	
4	3	4	

For more practice use Math Traveler.™

Name_____

Write the missing numbers.

1 | 645 | _____ | 647 | **2** | 398 | _____ | 400 |

Compare. Write >, <, or =.

3 545 ◯ 645   **4** 234 ◯ 234   **5** 333 ◯ 233

6 Write the numbers in order from greatest to least.

645 219 637 302

_____, _____, _____, _____

7 Write the numbers in order from least to greatest.

812 218 182 821

_____, _____, _____, _____

Write the missing number in each pattern.
Circle the pattern you used to count.

8 225, 325, 425, _____, 625, 725

hundreds   tens   ones

9 434, _____, 454, 464, 474, 484

hundreds   tens   ones

10 June's family takes a trip. They drive 100 miles a day. How far do they travel in 5 days?

_____ miles

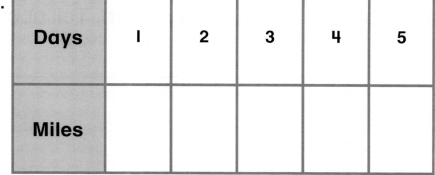

Days	1	2	3	4	5
Miles					

Assessment

Choose the best answer.

1 What is the time?

4:45

4:15

3:45

3:15

---

2 Which figure is a cylinder?

---

3 Hector has 27 stickers.
Kyra gives him 18 more stickers.
Which shows how many stickers Hector has now?

27 + 18   27 − 18   45 + 27   45 − 27

Solve.

4 Britney had $1.40. Then she found 3 nickels.
Now how much money does Britney have? $_____.

THINK SOLVE EXPLAIN

5 Make a skip-counting pattern. Tell about your pattern.

_____

3-Digit Addition

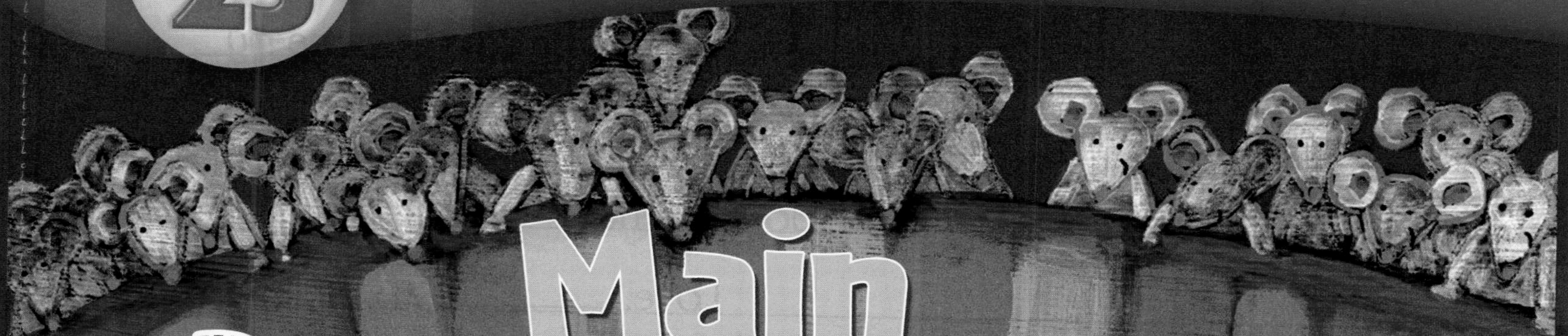

READ TOGETHER

Main Attraction

Everybody step right up—

Our show's a special treat!

A hundred acts, a million thrills—

So hurry, take your seat!

And in this ring, a superstar

As big as big can be.

Its feet are size six hundred ten,

And mine are just size three!

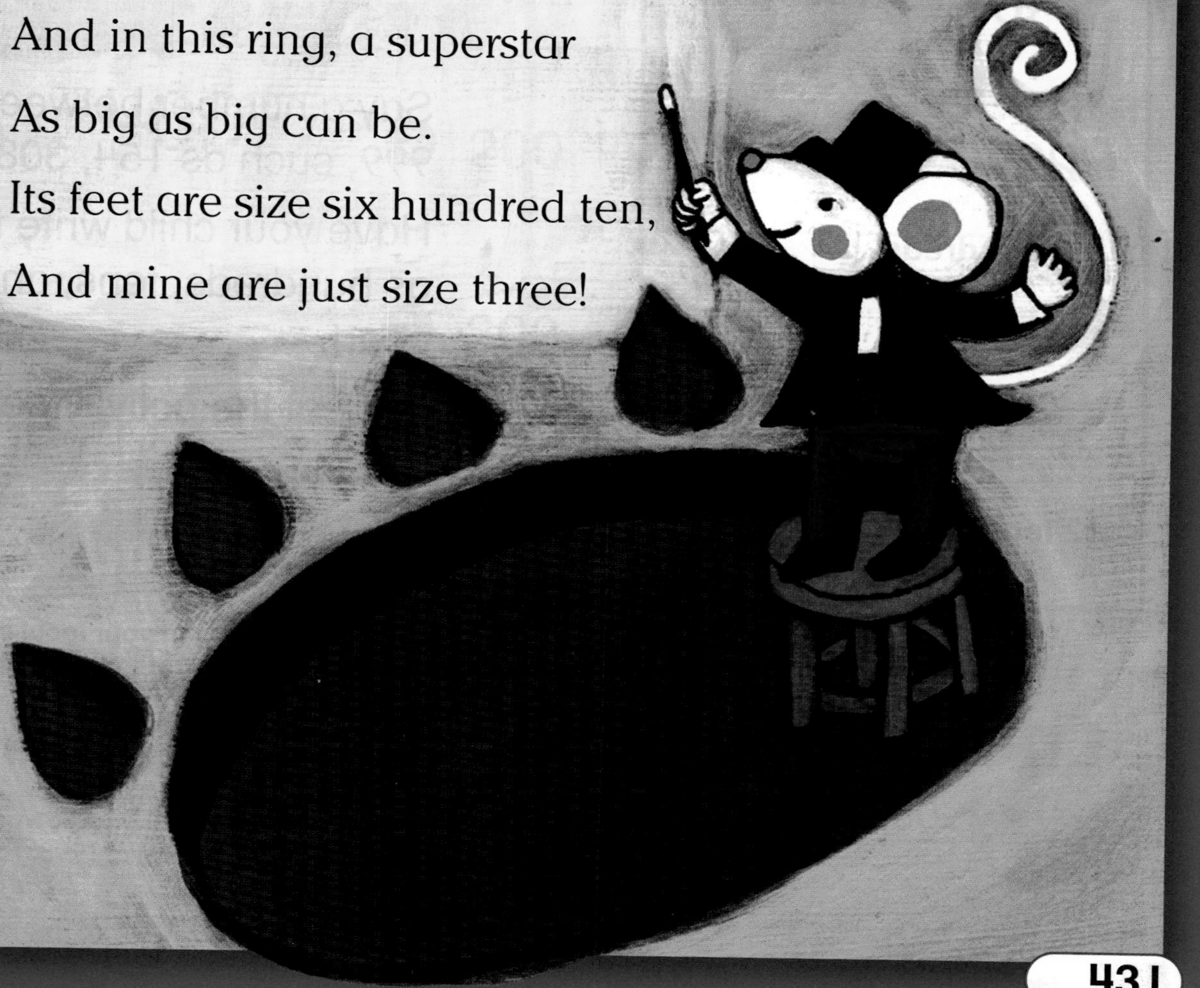

Practice   Add.

I know 4 + 2 = 6.
I use that to add
400 + 200.

10. 400 + 200 = 600

11. 300 + 400 = _____

12. 800 + 100 = _____

13. 700 + 200 = _____

14.
```
  300
+ 300
```

15.
```
  500
+ 300
```

16.
```
  500
+ 100
```

17.
```
  300
+ 200
```

18.
```
  100
+ 300
```

19.
```
  200
+ 600
```

20.
```
  400
+ 400
```

21.
```
  700
+ 200
```

22.
```
  200
+ 100
```

23.
```
  300
+ 400
```

24.
```
  100
+ 800
```

25.
```
  100
+ 600
```

Math at Home: Your child used basic facts to add hundreds.
Activity: Have your child show you how to add 200 + 100.

434   four hundred thirty-four

Name_____

Add. Color exercises where you regroup 🖍 .
Color other exercises 🖍 .

$$\begin{array}{r} 35¢ \\ +51¢ \\ \hline 86¢ \end{array}$$

¢ is a cent sign.

$$\begin{array}{r} 54¢ \\ +\ 8¢ \\ \hline \end{array}$$

$$\begin{array}{r} 16¢ \\ +27¢ \\ \hline \end{array}$$

$$\begin{array}{r} 53¢ \\ +44¢ \\ \hline \end{array}$$

$$\begin{array}{r} 28¢ \\ +53¢ \\ \hline \end{array}$$

$$\begin{array}{r} 40¢ \\ +39¢ \\ \hline \end{array}$$

$$\begin{array}{r} 42¢ \\ +38¢ \\ \hline \end{array}$$

$$\begin{array}{r} 67¢ \\ +\ 8¢ \\ \hline \end{array}$$

$$\begin{array}{r} 20¢ \\ +74¢ \\ \hline \end{array}$$

$$\begin{array}{r} 42¢ \\ +25¢ \\ \hline \end{array}$$

$$\begin{array}{r} 15¢ \\ +27¢ \\ \hline \end{array}$$

$$\begin{array}{r} 58¢ \\ +19¢ \\ \hline \end{array}$$

$$\begin{array}{r} 36¢ \\ +15¢ \\ \hline \end{array}$$

$$\begin{array}{r} 39¢ \\ +56¢ \\ \hline \end{array}$$

$$\begin{array}{r} 30¢ \\ +61¢ \\ \hline \end{array}$$

Practice Use 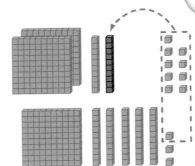, and ▪ to add.

If there are 10 or more ones, you need to regroup.

③

hundreds	tens	ones
2	⬚1	9
+ 1	5	3
3	7	2

④

hundreds	tens	ones
	⬚	
5	2	9
+ 2	3	7

hundreds	tens	ones
	⬚	
4	3	8
+ 3	1	2

hundreds	tens	ones
	⬚	
1	5	4
+ 2	4	3

⑤

hundreds	tens	ones
	⬚	
2	6	5
+ 1	2	7

hundreds	tens	ones
	⬚	
3	2	4
+ 2	6	5

hundreds	tens	ones
	⬚	
2	4	6
+ 2	1	9

Problem Solving ⟩ Number Sense

Show Your Work

Solve.

⑥ Mr. Song's class collects cans of food for charity. The class collects 118 cans the first week and 127 cans the second week. How many cans of food does the class collect in all?

_____ cans of food

Math at Home: Your child added 3-digit numbers.
Activity: Have your child draw models to show you how to add 137 + 219.

438 four hundred thirty-eight

Name_____

HANDS ON
Activity

Learn Use 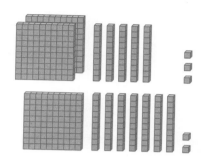, and
to find the sum of 253 and 172.

Add 253 and 172.

Step 1

Add the ones.
Write how many ones.

hundreds	tens	ones
□ 2	5	3
+ 1	7	2
		5

Step 2

Add the tens. If there are 10 or
more tens you need to regroup.
Regroup 10 tens as 1 hundred.
Write how many tens.
Write 1 to show the new hundred.

hundreds	tens	ones
□ 2	5	3
+ 1	7	2
	2	5

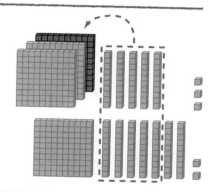

Step 3

Add the hundreds.
Write how many hundreds.

$253 + 172 = \underline{425}$

hundreds	tens	ones
1 2	5	3
+ 1	7	2
4	2	5

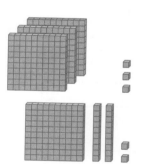

Your Turn Use , , and to add.

1.

hundreds	tens	ones
□ 3	6	4
+ 1	5	2
5	*1*	*6*

hundreds	tens	ones
□ 2	8	5
+ 2	4	3

hundreds	tens	ones
□ 1	5	2
+ 2	9	6

2. Write **About It!** Explain how you know when to regroup tens.

Practice

Use 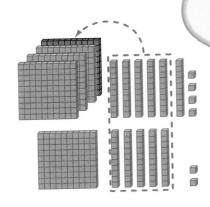, and
to add.

If there are 10 or more tens you need to regroup.

3

hundreds	tens	ones
3	6	4
+ 1	5	2
5	1	6

4

hundreds	tens	ones
3	7	2
+ 1	4	3

hundreds	tens	ones
2	9	1
+ 2	5	4

hundreds	tens	ones
1	6	4
+ 1	8	1

5

hundreds	tens	ones
1	4	3
+ 2	3	5

hundreds	tens	ones
2	4	6
+ 1	9	3

hundreds	tens	ones
3	4	2
+ 1	5	4

Make it Right

6 This is how Cora added 264 and 173.

Tell what mistake she made. Explain how to correct it.

```
  264
+ 173
-----
 3137
```

Math at Home: Your child added 3-digit numbers.
Activity: Have your child show you how to add 134 + 291 and explain the regrouping to you.

The Circus

The circus is in town. 175 people come to the show on Monday, 115 on Tuesday, and 104 on Wednesday. On Thursday there's no show. But 178 people come on Friday. And over the weekend, 694 people in all come to see the circus show!

 **Make Inferences**

1 Why do you think more people came to the show on the weekend?

_____

_____

2 How many people in all saw the circus show

on Monday and Tuesday? _____ people

3 How many people in all saw the circus show

on Wednesday and Friday? _____ people

4 How many people came to the show on Tuesday and Friday?

_____ people

Sunday at the Circus

Many people come to the circus on Sunday. They buy 167 bags of peanuts and 215 bags of popcorn. They buy 383 bottles of water and 231 bottles of soda.

 Reading Skill

Make Inferences

5 Why do you think people buy so much water or soda to drink?

_____

_____

6 How many bottles of water and soda

do they buy? _____ bottles

7 How many bags of peanuts and

popcorn do they buy? _____ bags

Math at Home: Your child made inferences to answer questions.
Activity: Have your child make an inference or judgment about how many people were at the circus on Sunday.

Problem Solving Practice

Solve.

1. Yesterday the ate 213 peanuts.

   Today the 🐘 ate 348 peanuts.

   How many 🥜 did it eat in all?

   213 + 348 = _____

2. Last year the 🎪 put on 290 shows.

   This year the 🎪 put on 313 shows.

   How many shows did the 🎪 put on in all?

   290 + 313 = _____

Write a Story!

THINK SOLVE EXPLAIN

3. Use the number sentence to write an addition problem about a circus. Find the sum.

   318 + 269 = _____

   _____

   _____

4. Last weekend the circus sold 342 posters. This weekend the circus sold 386 posters. How many posters did they sell in all?

   342 + 386 = _____

5. Last week the circus traveled 266 miles to get to our town. This week it traveled 314 miles to the next city. How many miles did it travel in all?

   266 + 314 = _____

Problem Solving

Writing for Math

Is Eric's estimate reasonable?

$108 + 294$ is about 400.

Writing

Think

I can find the nearest hundred for each number.

Solve

I know that 108 is between 100 and 200. It is closer to 100.

I know that 294 is between 200 and 300. It is closer to 300.

Then I estimate.

_____ + _____ = _____

Explain

I can tell you why the answer is reasonable.

_____

_____

_____

e-Journal **www.mmhmath.com**
Write about math

Name_____

Add.

1. 200 + 400 = _____

2. 100 + 700 = _____

3. 300 + 600 = _____

4. 500 + 200 = _____

Add. Regroup when you need to.

5.

hundreds	tens	ones
	☐	
1	4	9
+ 1	2	3

6.

hundreds	tens	ones
	☐	
2	7	1
+ 1	0	4

7.

hundreds	tens	ones
☐		
3	8	2
+ 1	7	5

8.

hundreds	tens	ones
☐		
3	7	6
+ 2	4	3

9. The elephants ate 150 peanuts for breakfast. They ate 250 peanuts for lunch. How many peanuts did they eat in all?

_____

10. The big elephant eats 400 peanuts for dinner. The little elephant eats 150 peanuts for dinner. Why do you think the little elephant eats fewer peanuts?

_____

Assessment

Spiral Review and Test Prep

Chapters 1—23

1 Choose the number for: 3 hundreds 6 ones.

36 ⭘   306 ⭘   360 ⭘   3006 ⭘

2 Which is 100 more than 428?

328 ⭘   429 ⭘   438 ⭘   528 ⭘

Use the table.

Bags of Snacks Sold		
Day	Popcorn	Peanuts
Saturday	245	176
Sunday	324	118

3 What was the total number of bags of peanuts sold on Saturday and Sunday?   _____ bags

4 How many bags of popcorn and peanuts did the circus sell on Sunday?   _____ bags

THINK SOLVE EXPLAIN

5 Use the numbers 2, 7, and 9. What is the greatest 3-digit number you can write?   _____

Explain your answer. _____

_____

_____

Test Prep

3-Digit Subtraction

SEA SONG

Sung to the tune of "My Bonnie Lies Over the Ocean"

I went to the beach on vacation,

To take a cool dip in the sea,

And when I went into the water,

A large school of fish swam by me.

I stopped and I started to count them—

I got up to 153.

But just then a wave knocked me over,

And swept all the fish out to sea!

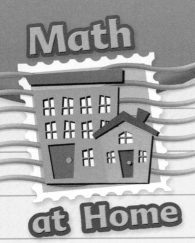

Math at Home

Dear Family,

I will learn how to subtract 3-digit numbers in Chapter 24. Here are my math words and an activity that we can do together.

Love, _____

My Math Words

difference :

$$\begin{array}{r} 436 \\ -\ 212 \\ \hline 224 \end{array}$$ ← difference

regroup :

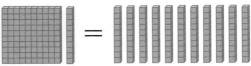

I hundred I ten = II tens

Home Activity

Work with your child to make up several subtraction problems involving 1- and 2-digit numbers. For example, subtract the number of letters in your name from 43.

Brian
$$\begin{array}{r} 43 \\ -\ 5 \\ \hline 38 \end{array}$$

Have your child solve the problems.

Books to Read

In addition to this library book, look for the **Time For Kids** math story that your child will bring home at the end of this unit.

- **Tightwad Tod** by Daphne Skinner, The Kane Press, 2001.
- **Time For Kids**

LOG ON

www.mmhmath.com
For Real World Math Activities

Name_____

Learn You can use subtraction facts to subtract hundreds.

Subtract $600 - 200$.

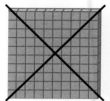 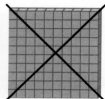

> I know that $6 - 2 = 4$.
> I use that to subtract hundreds.
> $600 - 200 = 400$

6 hundreds $- 2$ hundreds $= \underline{4}$ hundreds

$600 - 200 = \underline{400}$

Try It Subtract.

1 $500 - 100 = \underline{400}$

2 $900 - 200 = \underline{\hspace{1cm}}$

3 $700 - 400 = \underline{\hspace{1cm}}$

4 $600 - 100 = \underline{\hspace{1cm}}$

5 $800 - 500 = \underline{\hspace{1cm}}$

6 $400 - 300 = \underline{\hspace{1cm}}$

7 $300 - 100 = \underline{\hspace{1cm}}$

8 $200 - 100 = \underline{\hspace{1cm}}$

9 ✎ **Write About It!** What fact can you use to subtract $900 - 300$?

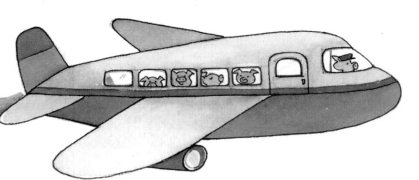

Practice Subtract.

10 700
−300
400

> Think:
> 7 − 3 = 4.
> So 700 − 300 = 400.

11 800
−100

12 400
−100

13 800
−400

14 500
−200

15 600
−400

16 900
−300

17 400
−200

18 300
−200

19 600
−300

20 700
−200

21 500
−400

22 800
−300

23 900
−100

24 500
−300

25 600
−500

26 700
−100

27 800
−600

Spiral Review and Test Prep

Choose the best answer.

28 Which number is 100 less than 400?

300 400 500 600
◯ ◯ ◯ ◯

29 Which would be the best unit to use to measure the length of a pencil?

inch foot yard meter
◯ ◯ ◯ ◯

 Math at Home: Your child used basic facts to subtract hundreds.
Activity: Have your child subtract 900 − 700.

Name_____

Subtract.

Sometimes you need to regroup to subtract.

$$\begin{array}{r} 72 \\ -42 \\ \hline \end{array}$$

$$\begin{array}{r} 61 \\ -\ 7 \\ \hline \end{array}$$

$$\begin{array}{r} 48 \\ -20 \\ \hline \end{array}$$

$$\begin{array}{r} 93 \\ -75 \\ \hline \end{array}$$

$$\begin{array}{r} 56 \\ -25 \\ \hline \end{array}$$

$$\begin{array}{r} 72 \\ -14 \\ \hline \end{array}$$

$$\begin{array}{r} 88 \\ -23 \\ \hline \end{array}$$

$$\begin{array}{r} 63 \\ -54 \\ \hline \end{array}$$

$$\begin{array}{r} 59 \\ -34 \\ \hline \end{array}$$

$$\begin{array}{r} 47 \\ -27 \\ \hline \end{array}$$

$$\begin{array}{r} 64 \\ -41 \\ \hline \end{array}$$

$$\begin{array}{r} 41 \\ -27 \\ \hline \end{array}$$

$$\begin{array}{r} 97 \\ -56 \\ \hline \end{array}$$

$$\begin{array}{r} 23 \\ -17 \\ \hline \end{array}$$

$$\begin{array}{r} 35 \\ -\ 8 \\ \hline \end{array}$$

$$\begin{array}{r} 81 \\ -39 \\ \hline \end{array}$$

$$\begin{array}{r} 32 \\ -15 \\ \hline \end{array}$$

Extra Practice

Draw a line or lines to make the new shapes.

1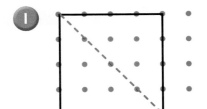

Make 2 triangles.

2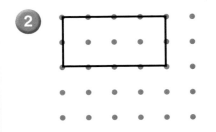

Make 2 squares.

3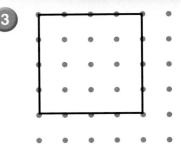

Make 4 squares.

4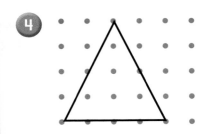

Make 2 triangles.

5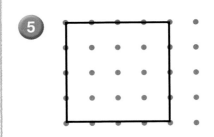

Make 2 rectangles.

6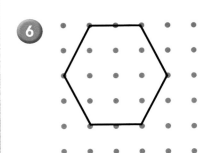

Make 6 triangles.

7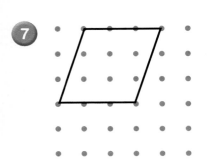

Make 3 triangles.

8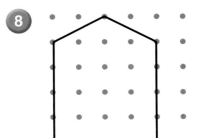

Make 1 triangle
and 1 square.

9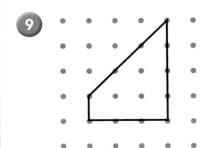

Make 1 triangle
and 1 rectangle.

Math at Home: Your child practiced making shapes.
Activity: Draw a square. Ask your child to draw lines to make 4 triangles.

Regroup Tens as Ones

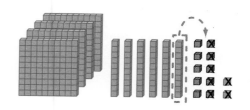

Learn You can use hundreds, tens, and ones models to subtract. Use a workmat and ▦, ▭▭, and ▪ to subtract 462 − 247.

Step 1

To subtract the ones you need to take away 7 ▪. You need to regroup 1 ten as 10 ones. Subtract the ones.

hundreds	tens	ones
4	5̸6	1̲2̲ 2̸
− 2	4	7
		5

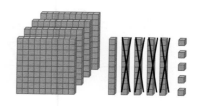

Step 2

To subtract the tens take away 4 ▭▭. Write how many tens are left.

hundreds	tens	ones
4	5 6̸	12 2̸
− 2	4	7
	1	5

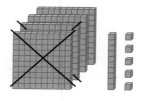

Step 3

To subtract the hundreds take away 2 ▦. Write how many hundreds are left.

hundreds	tens	ones
4	5 6̸	12 2̸
− 2	4	7
2	1	5

Your Turn Use ▦, ▭▭, and ▪ to subtract.

1.

hundreds	tens	ones
3	6 7̸	14 4̸
− 1	4	8
2	2	6

hundreds	tens	ones
2	8	3
− 1	3	5

hundreds	tens	ones
3	5	1
− 2	3	7

2. ✎ **Write About It!** How is subtracting 3-digit numbers like subtracting 2-digit numbers?

Practice Use 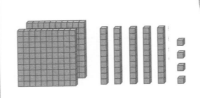, ▬, and ▫ to subtract.

If there are not enough ones to subtract you need to regroup.

3

hundreds	tens	ones
2	5̶ ⁴	4̶ ¹⁴
− 1	3	9
1	1	5

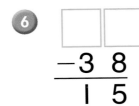

4

hundreds	tens ☐	ones ☐
4	8	5
− 1	5	7

hundreds	tens ☐	ones ☐
5	9	6
− 3	4	8

hundreds	tens ☐	ones ☐
3	4	7
− 2	1	3

5

hundreds	tens ☐	ones ☐
2	6	3
− 1	4	6

hundreds	tens ☐	ones ☐
6	7	4
− 2	2	5

hundreds	tens ☐	ones ☐
4	5	1
− 3	1	4

⭐ *x* **Algebra • Missing Numbers**

6
$$\begin{array}{r} \boxed{}\ \boxed{} \\ -\ 3\ 8 \\ \hline 1\ 5 \end{array}$$

7
$$\begin{array}{r} 6\ 1 \\ -\ \boxed{}\ \boxed{} \\ \hline 2\ 3 \end{array}$$

8
$$\begin{array}{r} \boxed{}\ \boxed{} \\ -\ 3\ 8 \\ \hline 9 \end{array}$$

 Math at Home: Your child subtracted 3-digit numbers by regrouping tens.
Activity: Have your child draw models to show you how to subtract 284 − 136.

Name_____

Regroup Hundreds as Tens

Learn Use a workmat and 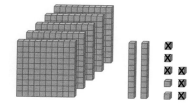, and ▫ to find the difference of 528 and 176.

Step 1

To subtract the ones take away 6 ▫. Write how many ones are left.

hundreds	tens	ones
□ 5	□ 2	8
− 1	7	6
		2

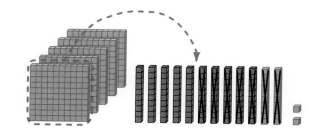

Step 2

To subtract the tens you need to take away 7 ▭. You need to regroup 1 hundred as 10 tens. Now subtract the tens.

hundreds	tens	ones
[4] 5̸	[12] 2̸	8
− 1	7	6
	5	2

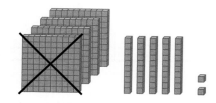

Step 3

Subtract the hundreds. Write how many hundreds are left.

hundreds	tens	ones
[4] 5̸	[12] 2̸	8
− 1	7	6
3	5	2

Your Turn Use , ▭, and ▫ to subtract.

1

hundreds	tens	ones
[2] 3̸	[11] 1̸	5
− 1	5	2
1	6	3

hundreds	tens	ones
□ 4	□ 5	7
− 1	8	5

hundreds	tens	ones
□ 5	□ 3	4
− 3	7	1

2 ✎ Write **About It!** Explain how you know when to regroup hundreds to subtract.

© Macmillan/McGraw-Hill

Practice   Use , ———, and ▫ to subtract.

> If there are not enough tens to subtract you need to regroup.

3

hundreds	tens	ones
²3̸	¹⁴4̸	9
− 1	7	5
1	7	4

4

hundreds	tens	ones
☐ 3	☐ 2	6
− 1	4	5

hundreds	tens	ones
☐ 5	☐ 3	7
− 2	9	4

hundreds	tens	ones
☐ 6	☐ 5	4
− 1	8	4

5

hundreds	tens	ones
☐ 4	☐ 9	7
− 2	6	5

hundreds	tens	ones
☐ 7	☐ 1	8
− 3	4	2

hundreds	tens	ones
☐ 5	☐ 4	6
− 1	9	3

Make it Right

THINK SOLVE EXPLAIN

6

hundreds	tens	ones
☐ 4	☐ 3	8
− 1	7	5
3	4	3

This is how Ben subtracted
438 − 175.
Why is Ben wrong? Make it right.

 Math at Home: Your child subtracted 3-digit numbers by regrouping hundreds.
Activity: Have your child show you how to subtract 438 − 251 and explain the regrouping to you.

456   four hundred fifty-six

Name_____

 Earl has $5.65. He wants to buy sunglasses for $3.70. How much money will he have left?

$3.70

Subtract to find out how much money is left.

$$\begin{array}{r} \overset{4\ \ 16}{\$5.\cancel{6}\cancel{5}} \\ -\ \ 3.70 \\ \hline \$1.95 \end{array}$$

You subtract money amounts like you subtract numbers.

Estimate to see if the answer is reasonable.

| $0.00 | $1.00 | $2.00 | $3.00 | $4.00 | $5.00 | $6.00 |

$3.70 $5.65

$$\begin{array}{r} \$5.65 \\ -\ \ 3.70 \\ \hline \$1.95 \end{array} \quad \begin{array}{l} \text{nearest dollar} \\ \text{nearest dollar} \\ \text{is about} \end{array} \quad \begin{array}{r} \$6.00 \\ -\ \ 4.00 \\ \hline \$2.00 \end{array}$$

Use the number line to help estimate.

$1.95 is close to $2.00.
The answer is reasonable.

 Try It Add or subtract. Estimate to see if your answer is reasonable.

1
$$\begin{array}{r} \overset{2\ \ 17}{\$4.\cancel{3}7} \\ -\ \ 2.18 \\ \hline \$2.19 \end{array} \quad \begin{array}{r} \$4.00 \\ -\ \ 2.00 \\ \hline \$2.00 \end{array}$$

2
$$\begin{array}{r} \$5.85 \\ +\ \ 2.69 \\ \hline \end{array} \quad +\ \underline{}$$

3 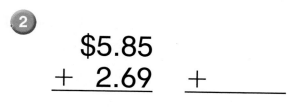 Write **About It!** How do you subtract two money amounts?

Practice Add or subtract. Estimate to see if your answer is reasonable.

> Rewrite the amounts to the nearest dollar to estimate.

```
←————|————|————|————|————|————|————|————|————→
   $0.00  $1.00  $2.00  $3.00  $4.00  $5.00  $6.00  $7.00  $8.00
```

4
```
   $2.17    nearest dollar →   $2.00
 +  3.26    nearest dollar → +  3.00
   $5.43       is about    →   $5.00
```

5
```
   $4.95
 +  2.13   + _____
```

6
```
   $5.09
 +  3.25   + _____
```

7
```
   $4.14
 +  1.78   + _____
```

8
```
   $6.74
 −  2.58   − _____
```

9
```
   $5.26
 −  3.41   − _____
```

10
```
   $7.28
 −  4.36   − _____
```

11
```
   $5.29
 −  2.95   − _____
```

Problem Solving **Critical Thinking**

12 Paco has $7.92. He wants to buy the boat for $4.75. He estimates he will still have enough money to buy the beach ball. Is Paco right? Explain your answer.

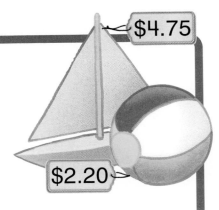

$4.75

$2.20

_____

_____

Math at Home: Your child subtracted and estimated money amounts.
Activity: Look in the newspaper for two items priced between $1.00 and $6.00. Have your child subtract the lesser price from the greater price.

Problem Solving Strategy

Work Backward

Sometimes you start with what you know and work backward to solve a problem.

The Wilson family is taking a 350 mile road trip. They have 218 miles left to go. How many miles did they already travel?

Problem Solving

Read

What do I already know?

The trip is _____ miles long.

They have _____ miles left to go.

What do I need to find out? _____

Plan

I can subtract the miles they have left to go. That will tell me how many miles they have already traveled.

Solve

I can carry out my plan.

The family has already traveled _____ miles.

$$\begin{array}{r} {}^{4}\;{}^{10} \\ 3\cancel{5}\cancel{0} \\ -218 \\ \hline 132 \end{array}$$

Look Back

How can I check my answer? _____

_____

Choose the best answer.

1 In the number 345, the 3 means

3	30	300	400
○	○	○	○

 2 What is the value of 400 + 30 + 6?

346	436	634	643
○	○	○	○

 3 Which number shows 10 more than 278?

268	278	288	378
○	○	○	○

4 Complete.

hundreds	tens	ones
☐	☐	☐
3	1	5
− 1	2	3

 5 The class collected 326 cans and 248 bottles. How many cans and bottles did the class collect?

_____ cans and bottles

 6 Write a number that comes between 431 and 440. Tell how you know.

_____

_____

Test Prep

D

TIME
FOR KIDS

Name _____

Juanita has a collection of 101 tiny toys.

Vicki has 108 tiny toys.
Who has more tiny toys?

How many more?

$$\begin{array}{r} 108 \\ -101 \\ \hline \end{array}$$

Fold down

TIME FOR KIDS

Counting Collections

People collect many different things.
See the collection of balls.

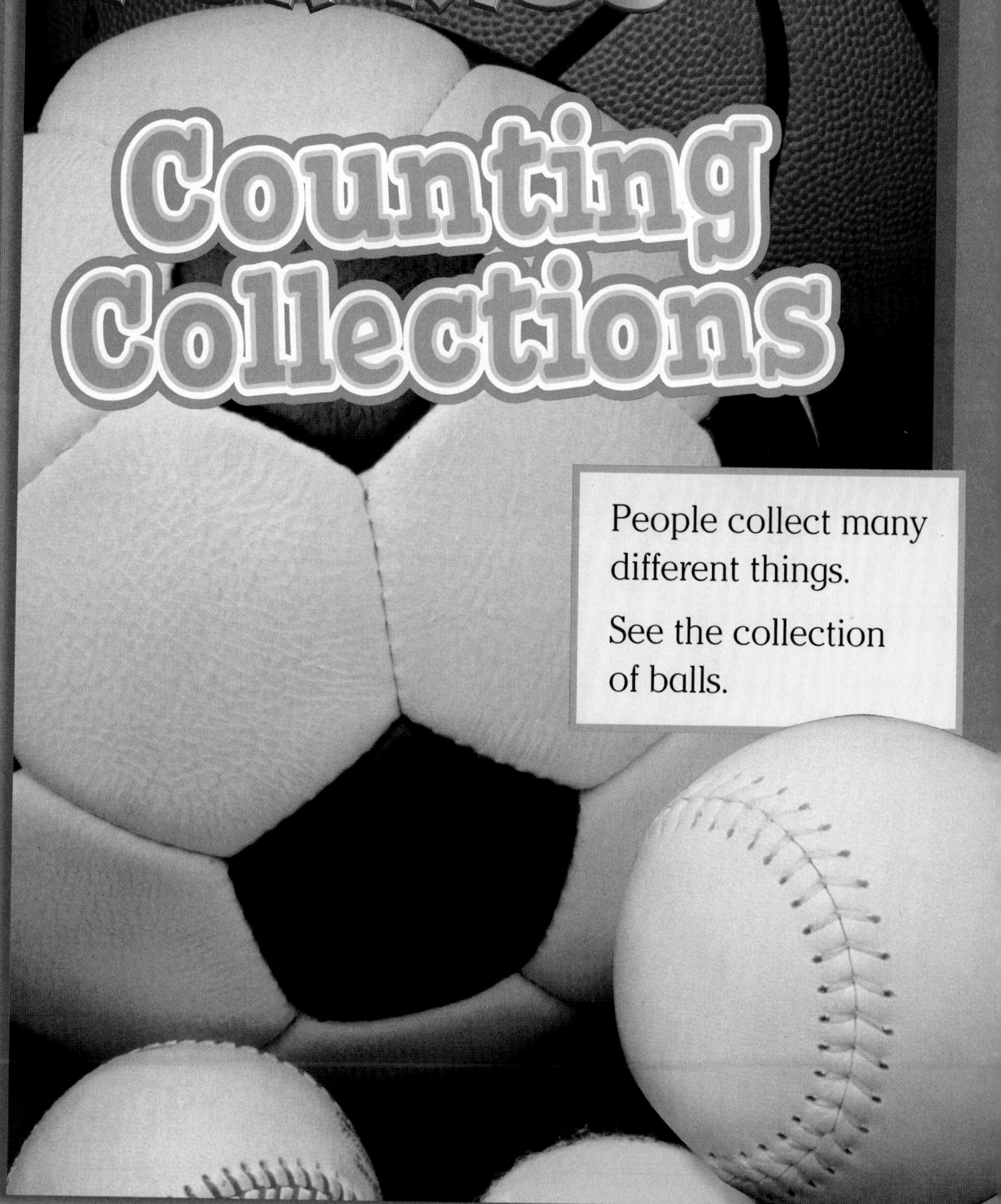

READ TOGETHER

You can count more than 100 seashells in this collection.

There are at least 125 stamps in this collection.

One jar has about 110 marbles.

There are about 250 coins in the other jar.

110 < 250

There are fewer marbles than coins.

Name_____

Addition and Water Use

Water comes from Earth. It is a natural resource we need to live. If we use too much water, we could run out of it. How could people save, or conserve, water?

Problem Solving

Water is a natural resource we need.

A leaky faucet uses up to 15 gallons of water each day.

Circle the word or words to complete each sentence.

1 Things that come from the Earth are _____.

natural resources   water

2 We must save or _____ water.

use up   conserve

© Macmillan/McGraw-Hill

What to Do

- Use  or ▦, ▭▭▭▭ , ▫.

- Choose one water use from the chart.

- Find how many gallons of water you use in a day. Write the total.

- Find how many gallons of water you use in a week. Write the total.

Water Use	Gallons Used
Flush toilet	5 gallons
Wash hands	1 gallon
Take bath	35 gallons
Take shower	25 gallons

Water Use	
Number of times in 1 day	
Gallons in 1 day	
Gallons in 1 week	

Problem Solving

Use your data to solve.

3 **Measure** How did you find the amount of water you used in one day?

_____

4 **Estimate** How could you estimate how much water you used in one week?

_____

_____

 5 **Compare** How does the amount of water you use compare with the amount your classmates use?

_____

 Math at Home: Your child applied addition to find how much water they use.
Activity: Repeat this activity. Choose a different water use from the chart. Have your child write the addition sentences for the daily use.

466 four hundred sixty-six

Name_____

Math Words

Draw lines to match.

1 =

2 >

3 <

| is greater than |
| is equal to |
| is less than |

Skills and Applications

Number Relationships and Patterns (pages 415-422)

Examples

Compare numbers.
Look at the hundreds first.
Then look at the tens and the ones.

320   >   234

320 is greater than 234.

4 821 ◯ 615

5 652 ◯ 693

6 901 ◯ 900

7 215 ◯ 330

Use number patterns to help you count.

Count by tens.

350, 360, 370, 380, 390

Count by hundreds.

418, 518, 618, 718, 818

8

| 610 | | 630 |
| 640 | 650 | |

9

| 299 | | 499 |
| 599 | | 799 |

Skills and Applications

Add and Subtract 3-Digit Numbers (pages 432-444; 448-460)

Examples

Add. Regroup if necessary.

hundreds	tens	ones
[1]	[1]	
2	7	7
+ 2	3	5
5	1	2

10

hundreds	tens	ones
[ ]	[ ]	
4	8	4
+ 3	6	7

Subtract. Regroup if necessary.

hundreds	tens	ones
[ ]	[6]	[15]
6	7̶	5̶
− 3	4	9
3	2	6

11

hundreds	tens	ones
[ ]	[ ]	[ ]
4	2	8
− 2	8	6

Problem Solving — Strategy

(pages 421–426)

Use a table to solve.

12 The craft shop orders 200 boxes of beads each week. How many boxes of beads does the shop order in 4 weeks? 5 weeks?

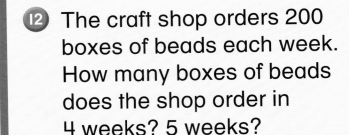

Weeks	Boxes
1	200
2	400
3	600
4	_____
5	_____

Math at Home: Your child learned about number relationships and patterns to 1,000 and adding and subtracting 3-digit aumbers.
Activity: Have your child use these pages to review number relationships and addition and subtraction of 3-digit numbers.

Name _____

Addition and Subtraction

Show as many combinations as you can.

335 695 529 205 148 179 285

Find 2 numbers with a sum less than 850. Show the sums.	Find 2 numbers with a difference less than 200. Show the differences.

_____ _____

_____ _____

_____ _____

_____ _____

_____ _____

_____ _____

Portfolio

You may want to put this page in your portfolio.

Unit 6   Performance Assessment
e-Journal   www.mmhmath.com
Write about math
four hundred sixty-nine   **469**

Assessment

Unit 6
Enrichment

Adding and Subtracting Money

Some friends went to the toy store.
Look at what they bought.
Then answer the questions.

A toy can be bought more than once.

Markers $1.97

Art Set $2.42

Checkers $2.56

BEAR $1.89

1. Paul bought 2 items totaling $4.53.
   Which 2 items did he buy?

_____

2. Megan bought 2 items totaling $3.86.
   Which 2 items did she buy?

_____

3. Samuel paid $4.00 for 1 item. He got
   $1.58 in change. Which item did he buy?

_____

4. Rachel paid $5.00 for 1 item. She got
   $2.44 in change. Which item did she buy?

_____

Fractions

READ TOGETHER

Under the Sea

Story by Susan Banta

Illustrated by Christine Mau

How many sea creatures swim with me?

_____ sea creatures

Only _____ of them are yellow fish.

How many starfish swim with me?

____ starfish

Only ____ starfish is hiding.

How many fish swim with me?

_____ fish

Only _____ of the fish are purple.

How many dolphins swim with me?

_____ dolphins

_____ of the dolphins are light gray.

Math at Home

Dear Family,

I will learn about fractions in Chapter 25. Here are my math words and an activity that we can do together.

Love, _____

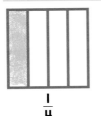

My Math Words

fraction :

a number that names part of a whole or a group

$\frac{1}{8}$ $\frac{1}{6}$ $\frac{1}{4}$ $\frac{1}{3}$ $\frac{1}{2}$

is greater than (>) :

$\frac{1}{2}$ > $\frac{1}{4}$

is less than (<) :

$\frac{1}{4}$ < $\frac{1}{2}$

LOG ON

www.mmhmath.com
For Real World Math Activities

Home Activity

Cut two shapes out of paper, such as a rectangle and a square.

Ask your child to fold each shape to make two or more equal parts.

Then have your child tell how many equal parts each shape has.

© Macmillan/McGraw-Hill

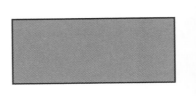

Books to Read

Look for these books at your local library and use them to help your child learn fractions.

- **Clean–Sweep Campers** by Lucille Recht Penner, The Kane Press, 2000.
- **Two Greedy Bears** by Mirra Ginsburg, Simon & Schuster, 1998.
- **Jump, Kangaroo, Jump!** by Stuart J. Murphy, HarperCollins, 1999.

Learn A fraction can name a part of a whole.
Use a fraction to name equal parts.

Math Words
fraction
equal parts

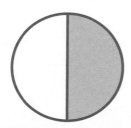

I of 2 equal parts is pink.
One half is pink.

$\frac{1}{2}$ $\frac{1 \text{ pink part}}{2 \text{ equal parts}}$

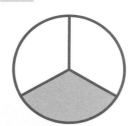

I of 3 equal parts is pink.
One third is pink.

$\frac{1}{3}$ $\frac{1 \text{ pink part}}{3 \text{ equal parts}}$

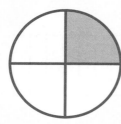

I of 4 equal parts is pink.
One fourth is pink.

$\frac{1}{4}$ $\frac{1 \text{ pink part}}{4 \text{ equal parts}}$

I of 8 equal parts is pink.
One eighth is pink.

$\frac{1}{8}$ $\frac{1 \text{ pink part}}{8 \text{ equal parts}}$

Try It Write the fraction.

 1

$\frac{\boxed{1}}{\boxed{4}}$ number of shaded parts

number of equal parts

 2

$\frac{\boxed{}}{\boxed{}}$ number of shaded parts

number of equal parts

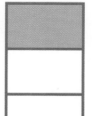

3 **Write About It!** How can you tell if the parts are equal?

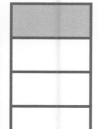

Practice Write the fraction for the shaded part.

> Write the number of shaded parts on top. Write the number of equal parts on the bottom.

4   $\frac{1}{2}$

5   _____

Color part of each shape to show the fraction.

6   $\frac{1}{3}$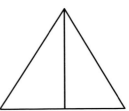

7   $\frac{1}{8}$

8   $\frac{1}{4}$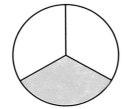

9   $\frac{1}{2}$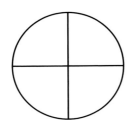

Make it Right

THINK
SOLVE
EXPLAIN

10   Dave says $\frac{1}{2}$ of the square is red.

Why is Dave wrong? Make it right.

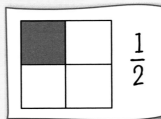 $\frac{1}{2}$

Math at Home: Your child learned about fractions as parts of a whole.
Activity: With your child, fold a sheet of paper into four equal parts. Have your child shade one part and then write a fraction to tell about his or her picture.

Name_____

Learn You can write a fraction for the whole.

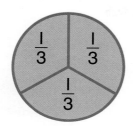

$\frac{1}{3}$ $\frac{1}{3}$ $\frac{1}{3}$

There are 3 green parts.
There are 3 equal parts.

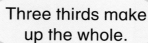
Three thirds make up the whole.

$\frac{3}{3}$ $\frac{\text{3 green parts}}{\text{3 equal parts}}$

The fraction for the whole is $\frac{3}{3}$.

The fraction for the whole always equals 1.

$$\frac{3}{3} = 1$$

Try It Count the parts in each whole.
Then write the fraction for the whole.

1  $\frac{2}{2}$

2 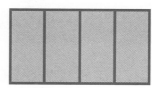 _____

3  _____

4  _____

5 ✎ Write **About It!** Why does a fraction for a whole have the same number on the top and the bottom?

© Macmillan/McGraw-Hill

Practice Count the parts in each whole.
Color the parts .
Then write the fraction for the whole.

The same whole can be made up of different numbers of parts.

6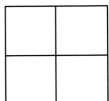

 $\frac{8}{8}$ is the same as 1.

$\frac{8}{8}$
___

7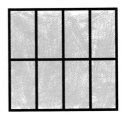

___

8

___

9

___

10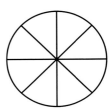

___

11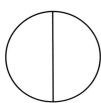

___

12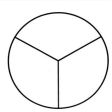

___

Problem Solving **Reasoning**

13 Each pizza shows halves.
Show how you count the halves.

$\frac{2}{2}$ $\frac{4}{2}$ $\frac{}{2}$ $\frac{}{2}$

 Math at Home: Your child learned about fractions for the whole.
Activity: After you cut a sandwich in equal halves or fourths for your child, ask your child to name the fraction for the whole. (2 or 4)

Name_____

 These fractions name more than one equal part.

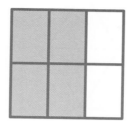
Three fourths and four sixths are fractions that name more than one equal part.

3 of 4 equal parts are blue.
Three fourths is blue.

$\frac{3}{4}$ $\frac{3 \text{ blue parts}}{4 \text{ equal parts}}$

4 of 6 equal parts are blue.
Four sixths is blue.

$\frac{4}{6}$ $\frac{4 \text{ blue parts}}{6 \text{ equal parts}}$

Try It Write the fraction for the shaded part.

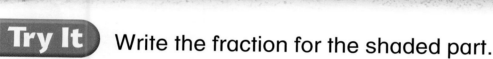

1.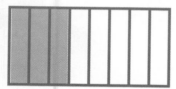

$\frac{3}{8}$ _____

2.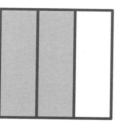

_____

3.

_____

4.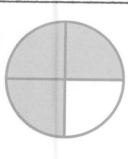

_____

5.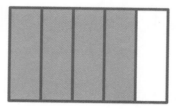

_____

6.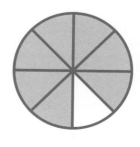

_____

7. ✏️ Write **About It!** How do you show $\frac{7}{10}$ of a rectangle?

Practice Color to show the fraction.

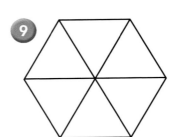

Remember, the bottom number is the number of parts in all.

8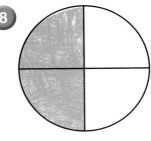

$\frac{2}{4}$

9

$\frac{3}{6}$

10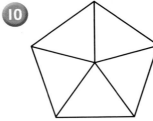

$\frac{4}{5}$

11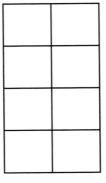

$\frac{5}{8}$

12

$\frac{2}{3}$

13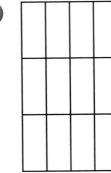

$\frac{4}{12}$

Problem Solving **Visual Thinking**

14 Color to show each fraction.

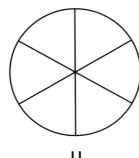

$\frac{2}{4}$ $\frac{4}{6}$

15 Which fraction shows more? Use your shaded circles to decide.

_____

 Math at Home: Your child learned about fractions.
Activity: Point to one or more shapes on this page that your child has shaded. Have him or her explain his/her work.

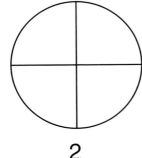

Name_____

Unit Fractions of a Group

HANDS ON
Activity

Learn Fractions can name one or more than one equal part of a group.

1 of 4 equal parts is red.

One fourth is red.

$\frac{1}{4}$ $\frac{1 \text{ red part}}{4 \text{ equal parts}}$

The top number tells how many parts you are talking about. The bottom number tells how many equal parts are in the group.

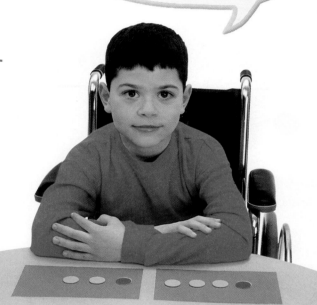

1 of 3 equal parts is red.

One third is red.

$\frac{1}{3}$ $\frac{1 \text{ red part}}{3 \text{ equal parts}}$

Try It Use . Write the fraction for the red part.

1

$\frac{1}{6}$ _____

2
_____

3

_____

4 ✏️ **Write About It!** What does the fraction $\frac{1}{5}$ mean?

Practice Write the fraction for the red part.

You can use counters to show the fraction.

5

$\frac{1}{8}$

6

———

7

———

8

———

Problem Solving — Number Sense

 Show Your Work

9 Draw a picture to explain.

8 turtles are on a rock. Half of them swim away. How many turtles are still on the rock?

_____ turtles

 Math at Home: Your child learned about fractions as parts of a group.
Activity: Gather 8 pennies. Turn 7 pennies to show heads and 1 penny to show tails. Ask your child to tell what fraction of the group shows heads. Continue with other amounts.

Name_____

Learn A fraction can name part of a group.
What fraction of each group of starfish is blue?

There are 2 equal parts.

1 of 2 equal parts is blue.

The blue part is $\frac{1}{2}$.

There are 4 equal parts.

3 of 4 equal parts are blue.

The blue part is $\frac{3}{4}$.

Try It Write the fraction for the blue part.
If you need help, use counters.

1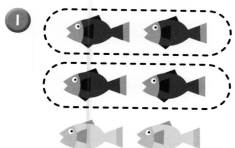

$\frac{2}{3}$

2

_____

3

_____

4

_____

5 ✎ Write **About It!** How would you use a picture to show
that $\frac{2}{3}$ of the group is blue?

Color the circled parts. Write a fraction for those parts.

6

$\frac{1}{2}$

7

_____

8

_____

9

_____

Problem Solving **Number Sense**

10 Circle the cards that show the same fraction.

$\frac{2}{3}$

two thirds

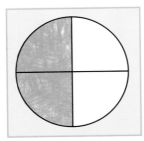

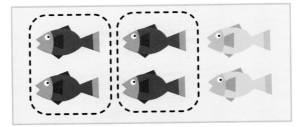

 Math at Home: Your child learned about fractions as parts of a group.
Activity: Invite your child to divide a group of 12 objects into halves, thirds, and then fourths.

Name_____

Compare Fractions

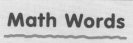

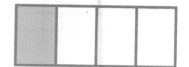 **Learn** You can compare fractions by using
is greater than > or **is less than <**.

$\frac{1}{4}$ is shaded.

$\frac{1}{4}$ is less than $\frac{1}{2}$.

$\frac{1}{4}$ ⬭< $\frac{1}{2}$

$\frac{1}{2}$ is shaded.

$\frac{1}{2}$ is greater than $\frac{1}{4}$.

$\frac{1}{2}$ ⬭> $\frac{1}{4}$

Math Words

is greater than >
is less than <

One half is
greater than
one fourth.

Try It Compare the shaded parts.
Then circle the fraction that is greater.

 1

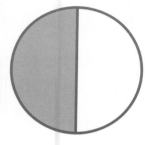

$\boxed{\frac{1}{2}}$ $\frac{1}{3}$

 2

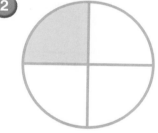

$\frac{1}{4}$ $\frac{1}{8}$

 3

 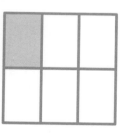

$\frac{1}{8}$ $\frac{1}{6}$

 4

 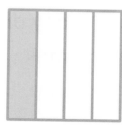

$\frac{1}{3}$ $\frac{1}{4}$

 5 Write **About It!** Which fraction is greater, $\frac{1}{4}$ or $\frac{1}{3}$? Explain.

© Macmillan/McGraw-Hill

Practice

Compare the shaded parts.
Then write < or >.

6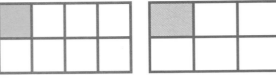

$\dfrac{1}{8}$ ⟨ < ⟩ $\dfrac{1}{6}$

7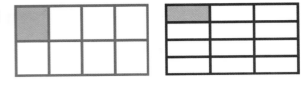

$\dfrac{1}{8}$ ◯ $\dfrac{1}{12}$

8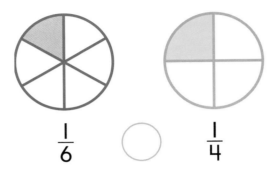

$\dfrac{1}{6}$ ◯ $\dfrac{1}{4}$

9

$\dfrac{1}{2}$ ◯ $\dfrac{1}{3}$

Compare the fractions. You may draw a picture.
Write < or >.

10 $\dfrac{1}{2}$ ⟨ > ⟩ $\dfrac{1}{4}$

11 $\dfrac{1}{3}$ ◯ $\dfrac{1}{8}$

12 $\dfrac{1}{5}$ ◯ $\dfrac{1}{4}$

13 $\dfrac{1}{4}$ ◯ $\dfrac{1}{8}$

14 $\dfrac{1}{12}$ ◯ $\dfrac{1}{6}$

15 $\dfrac{1}{4}$ ◯ $\dfrac{1}{3}$

Problem Solving — Visual Thinking

THINK
SOLVE
EXPLAIN

16 Tony drank $\dfrac{1}{6}$ of a glass of juice. Emily drank $\dfrac{1}{4}$ of a glass of juice. Tony drank more. Explain why.

Tony   Emily

_____

Math at Home: Your child compared fractions.
Activity: Divide one sheet of paper into 2 equal parts. Divide another sheet of paper into 4 equal parts. Ask your child what fraction each part shows. Ask him or her to tell you which fraction is greater.

484 four hundred eighty-four

Name_____

Picnic at the Beach

Jen, Brad, Mom, and Dad are at the beach. It's lunch time and everyone gets a sandwich to eat. Mom eats $\frac{1}{4}$ of her sandwich. Jen eats $\frac{1}{2}$ of hers.

Problem Solving

Reading Skill **Draw Conclusions**

1. Who eats more of her sandwich—Mom or Jen? Explain.

_____

2. How much of Mom's sandwich is left? _____

   How much of Jen's sandwich is left? _____

3. Brad has $\frac{2}{3}$ of a cup of lemonade. Jen has $\frac{1}{3}$ of a cup of lemonade.

   Who has more? _____

   How much more? _____ cup

© Macmillan/McGraw-Hill

Chapter 25 Lesson 7

Fun at the Beach

After lunch, Jen and Dad went swimming. Jen counted 6 waves. Brad and Mom walked along the beach. Mom and Brad collected 12 seashells. Later they got together again and talked about what they did.

Problem Solving

 Draw Conclusions

4 What part of the family went for a swim? Explain how you know.

_____

_____

5 2 of the waves Jen counted were very high. Write the fraction for the high waves. _____

6 Brad found 8 of the shells. Mom found the rest. Write the fraction for the shells Brad found. _____

Write the fraction for the shells Mom found. _____

 Math at Home: Your child was able to draw conclusions to answer questions.
Activity: Have your child read question 5 again and write a fraction for the waves that were not high.

486 four hundred eighty-six

Name_____

Problem Solving Practice

Solve.

① There are 3 crabs. One is blue. What fraction is blue? Circle.

$$\frac{1}{2} \qquad \frac{1}{3} \qquad \frac{2}{3}$$

② 4 🐟 swim in the sea.

4 🐟 swim away.

Circle the fraction that shows what part of the 🐟 swim away.

$$\frac{1}{3} \qquad \frac{5}{8} \qquad \frac{4}{4}$$

③ 6 divers look at fish. 4 of the divers are men. Write the fraction for the part of the divers that are men.

_____

What fraction of the divers are women?

_____

④ Jen and Dad caught 12 fish all together. Jen caught 5 of them. Write the fraction for the fish Jen caught.

_____

What fraction of the fish did Dad catch?

_____

THINK SOLVE EXPLAIN

Write a Story!

⑤ Write a problem about a group of 8 sea turtles. Use the fraction $\frac{5}{8}$.

_____

_____

© Macmillan/McGraw-Hill

Problem Solving

Writing for Math

THINK
SOLVE
EXPLAIN

Jason, Tony, and Mary shared a pizza.
Jason ate $\frac{1}{4}$ of the pizza. Tony ate $\frac{1}{2}$.
Mary ate $\frac{1}{8}$. Who ate the most?

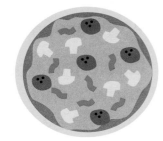

Writing

Think

The greatest fraction names the largest part.
I need to find the greatest fraction.
I can fold a paper circle to show fractions.

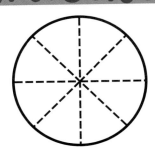

Solve

Color the paper circle to show each fraction.
Then I write the fractions in order from greatest to least.

_____

Explain

I can tell you who ate the most.

_____

_____

_____

Name_____

Write the fraction for the shaded part.

  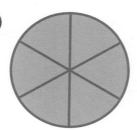 

1. _____   2. _____   3. _____

Color the circled parts. Write the fraction.

4. _____   5. _____   6. _____

Compare the fractions. Write < or >.

7. $\frac{1}{2}$ ◯ $\frac{1}{4}$ 8. $\frac{1}{3}$ ◯ $\frac{1}{2}$ 9. $\frac{1}{5}$ ◯ $\frac{1}{3}$

10. Kelly ate $\frac{1}{2}$ of a sandwich.

George ate $\frac{1}{2}$ of a sandwich.

Kelly ate more. Explain.

Kelly

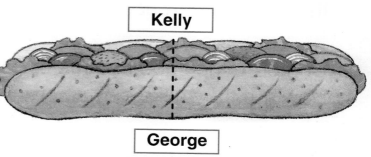 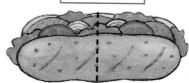
George

_____

_____

Assessment

Choose the best answer.

1 Which object may be 6 feet long?

○   ○   ○   ○

2 Add to find the sum.

$36 + 15 =$

51   61   71   81

○   ○   ○   ○

3 What fraction of the group is shaded?

$\frac{1}{2}$ $\frac{1}{3}$ $\frac{1}{4}$ $\frac{1}{5}$

○   ○   ○   ○

4 Write a subtraction sentence that has a difference of 20. _____

THINK SOLVE EXPLAIN

5 Compare the cube and the sphere. How are they alike and different?

_____

_____

Test Prep

Probability

READ TOGETHER

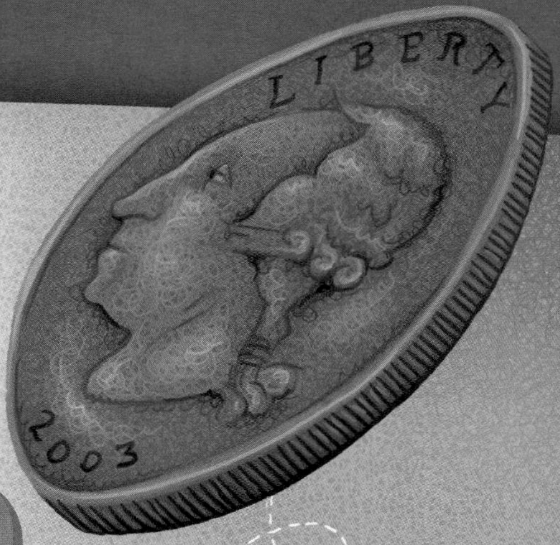

Heads or Tails?

I toss a quarter in the air

To see which side will show.

Sometimes heads, and sometimes tails —

I never seem to know!

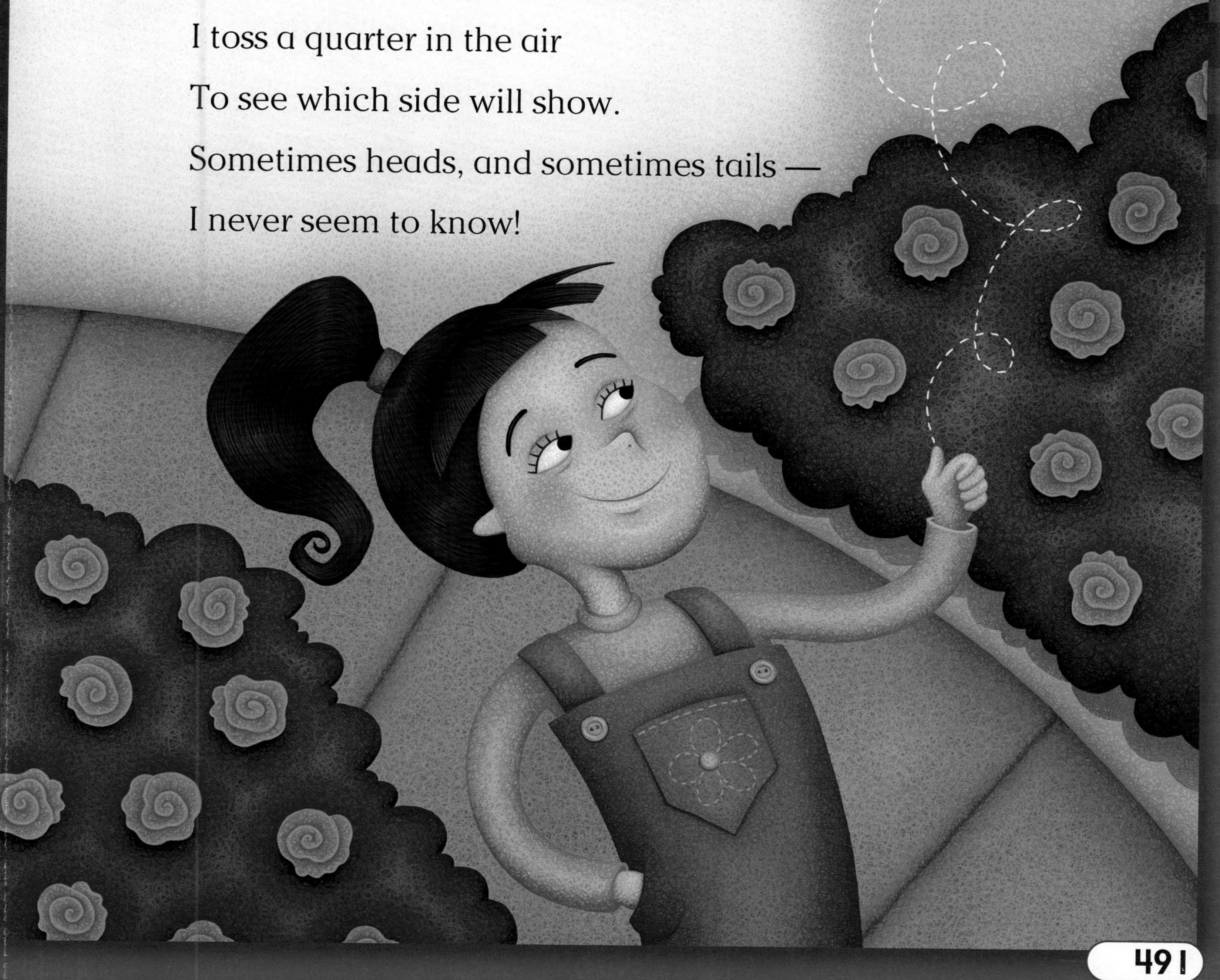

Math at Home

Dear Family,

I will learn about probability in Chapter 26. Here are my math words and an activity that we can do together.

Love, _____

My Math Words

equally likely :
same chance to happen

less likely :
not as probable

more likely :
probable

prediction :
telling before an event happens

I will spin red again.

www.mmhmath.com
For Real World Math Activities

Home Activity

Put 12 pennies on the table.

Have your child separate them into 2 equal groups.

Ask how many pennies there are in each group and what fraction each group represents.

Repeat the activity with 3, 4, and 6 equal groups.

© Macmillan/McGraw-Hill

Books to Read

Look for these books at your local library and use them to help your child learn probability.

- **Bad Luck Brad** by Gail Herman, The Kane Press, 2002.
- **No Fair!** by Caren Holtzman, Scholastic, 1997.
- **If You Give a Pig a Pancake** by Laura Numeroff, HarperCollins, 2000.

IF YOU GIVE A PIG A PANCAKE
BY Laura Numeroff
ILLUSTRATED BY Felicia Bond

Name_____

Learn You can tell if an event is certain , probable , or impossible .

Math Words
certain
probable
impossible

Certain	**Probable**	**Impossible**
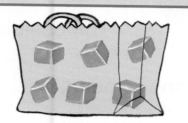		
Picking an orange cube from this bag is certain. It will always happen.	Picking an orange cube from this bag is probable. It is likely to happen.	Picking an orange cube from this bag is impossible. It will never happen.

Try It Circle the answer.

1 Picking an orange cube is

(certain)

probable

impossible

2 Picking a yellow cube is

certain

probable

impossible

3 Picking a yellow cube is

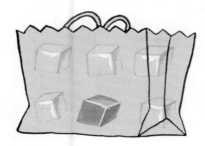

certain

probable

impossible

4 Picking an orange cube is

certain

probable

impossible

5 Write **About It!** Tell the colors of 8 cubes if picking orange is certain.

© Macmillan/McGraw-Hill

Practice Look at each spinner to answer the question. Circle the answer.

6 The spinner landing on blue is

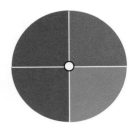

certain

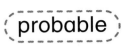
probable

impossible

7 The spinner landing on red is

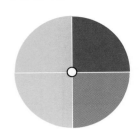

certain

probable

impossible

8 The spinner landing on yellow is

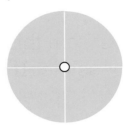

certain

probable

impossible

9 The spinner landing on green is

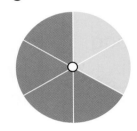

certain

probable

impossible

10 The spinner landing on blue is

certain

probable

impossible

11 The spinner landing on green is

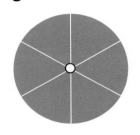

certain

probable

impossible

Problem Solving **Visual Thinking**

 Use the picture to answer the question.

12 Which color is impossible to pick? Explain your answer.

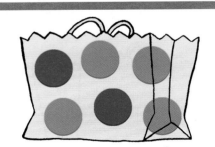

_____

 Math at Home: Your child discussed the likelihood of a given event using the words *certain*, *probable*, and *impossible*.
Activity: Take a plastic bag. Put 5 pennies inside. Ask your child to use the word *certain*, *probable*, or *impossible* that he or she would pick a penny from the bag. (certain) A dime? (impossible)

Name_____

HANDS ON Activity

**Learn** Sometimes an event is more likely , equally likely , or less likely to happen.

Toby is about to pick a triangle without looking.

Math Words
more likely
equally likely
less likely

Which color triangle is Toby more likely to pick?

Which is he less likely to pick?

What if there are the same number of and ?

Toby is ___equally likely___ to pick  or .

Your Turn

1 Look at each picture.
Put cubes in a bag.

- Which color are you more likely to pick? Color.

- Which color are you less likely to pick? Color.

- Then pick one cube without looking. Color the cube.

Bag	More Likely	Less Likely	Your Pick

2 ✏ **Write About It!** If there were 4 ▲ and 2 ▲, would Toby more likely pick a ▲ or ▲? Why?

Practice

3 Look at each picture.

Put cubes in a bag.

- Which color are you more likely to pick? Color.

- Which color are you less likely to pick? Color.

- Then pick one cube without looking. Color the cube.

Bag	More Likely	Less Likely	Your Pick

Problem Solving — Visual Thinking

THINK
SOLVE
EXPLAIN

4 Color the spinner using red 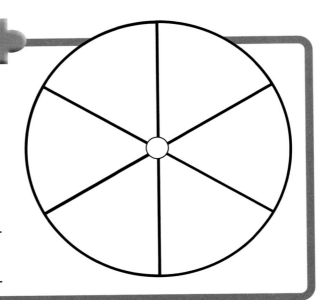 and blue 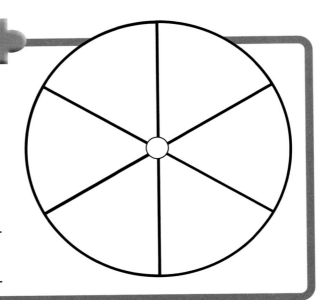 so that red or blue are equally likely to be spun. Tell how you decide.

_____

_____

Name_____ **Make Predictions**

Learn You can make a prediction that something will happen.

I predict you will spin yellow again.

Math Word

prediction

Your Turn Use red and yellow to make your own spinner. Then answer the questions.

1 What color do you predict you will spin more often? _____

2 Predict. If you spin the spinner 12 times, how many times will you get red? _____

Spin the spinner 12 times. Record each spin.

3 How many times did the spinner land on yellow? _____

4 How many times did the spinner land on red? _____

red	yellow

5 ✏ Write **About It!** Will your predictions always match what happens?

© Macmillan/McGraw-Hill

Practice Use a penny.
Answer the questions.

6 Predict. If you toss the penny 10 times, how many times will you get

 ? _____   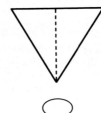 ? _____

Toss the penny 10 times. Record each toss.

7 How many times did you land on ? _____   ? _____

	1	2	3	4	5	6	7	8	9	10

Spiral Review and Test Prep

Choose the best answer.

8 Which picture shows a pie divided in fourths?

○   ○   ○   ○

9 Which picture shows a line of symmetry?

○   ○   ○   ○

Math at Home: Your child predicted outcomes.
Activity: Toss a nickel and have your child record how it landed. Repeat five times. Ask your child to predict how it will land next.

Name_____

Add or subtract.

42 +22	27 +25	49 +18	24 −15	73 −18
45 −21	62 +15	59 −19	32 +12	15 +19

© Macmillan/McGraw-Hill

Extra Practice

Color in the quilt to complete the pattern.

  Write About It! Draw a pattern using a circle, square, and triangle. Repeat it.

 www.mmhmath.com
For more Practice

 Math at Home: Your child practiced finding and extending patterns.
Activity: Draw a simple green and blue repeating pattern. Ask your child to draw what would most likely come next.

Problem Solving Strategy

Name_____

Make a List

Sometimes you need to organize information.

Tracey is making a game with 2-digit numbers. She uses the digits 3, 4, and 6 to make different 2-digit numbers. How many different numbers can she make?

Read

What do I already know? _____

What do I need to find out? _____

Plan

I need to find how many different numbers Tracey can make. I will make a list.

> Use each digit only once in the 2-digit number.

Solve

I can carry out my plan.

The list shows how many different 2-digit numbers.

There are ___6___ numbers.

Digits	2-Digit Numbers	
3	34	36
4	43	46
6	63	64

Look Back

Did I list all the numbers? _____

© Macmillan/McGraw-Hill

Problem Solving

Make a list to solve.

1 Emily uses the digits 2, 5, and 7 to make different 2-digit numbers. What numbers can she make?

Digits	2-Digit Numbers
2	____ ____
5	____ ____
7	____ ____

2 Trebor uses the digits 2, 4, and 8 to make different 2-digit numbers. What numbers can he make?

Digits	2-Digit Numbers
2	____ ____
4	____ ____
8	____ ____

3 Tracey's brother made different 2-digit numbers with 1, 5, and 9. What numbers can he make?

Digits	2-Digit Numbers
1	____ ____
5	____ ____
9	____ ____

4 Jason made different 2-digit numbers with 1, 2, 3, and 4. What numbers can he make?

Digits	2-Digit Numbers
1	____ ____ ____
2	____ ____ ____
3	____ ____ ____
4	____ ____ ____

Problem Solving

Math at Home: Your child learned to solve problems by making an organized list.
Activity: Have your child make a list to show all the different 2-digit numbers he or she can make using 3, 7, and 9.

Name_____

Name a Fraction!

 **2 players**

▶ Take turns. Put the cubes in the bag. Shake the bag and then take out 4 cubes.

▶ Put a ● over the fraction that names the red part.

▶ Put the cubes back in the bag.

▶ Play until both players complete the chart.

You Will Need

3 🔲

5 🔲

8 ●

a paper bag

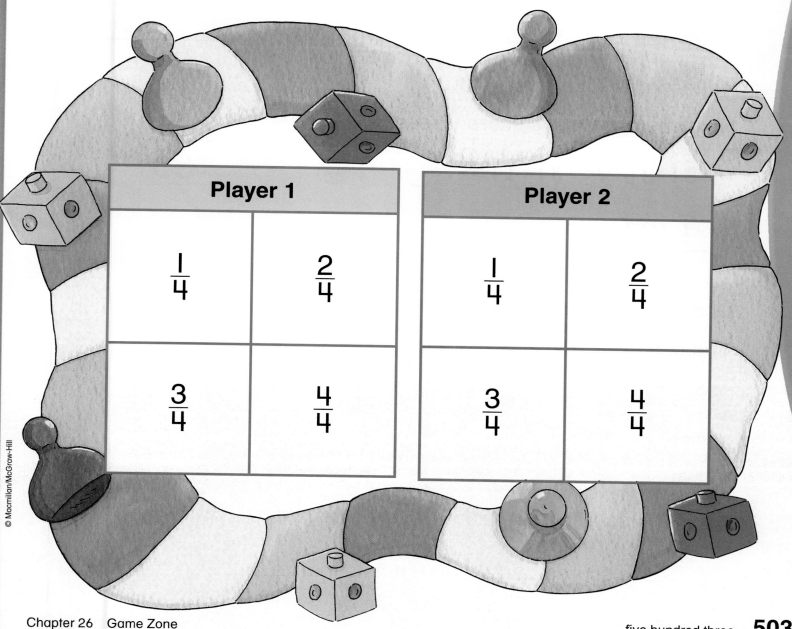

Player 1		Player 2	
$\frac{1}{4}$	$\frac{2}{4}$	$\frac{1}{4}$	$\frac{2}{4}$
$\frac{3}{4}$	$\frac{4}{4}$	$\frac{3}{4}$	$\frac{4}{4}$

Technology Link

Compare Data • Calculator

You can use a to compare data.

Tim plays 3 games with Jake.

	Game 1	Game 2	Game 3
Tim	23	45	17
Jake	38	31	14

Who scored more points? How many more?

First you add Tim's score. Press.

       **85**

Record it. Then press .

Next you add Jake's score. Press.

            83

_____ Tim _____ scored _____ 2 _____ more points than _____ Jake _____.

> 85 > 83;
> 85 − 83 = 2

Use data from the chart to solve.

You can use a .

	Game 1	Game 2	Game 3	Game 4
Amy	32	16	17	25
Kim	29	13	24	27

1 Who scored more points? How many more?

_____ scored _____ more points than _____ .

Name_____

Circle the answer.

1 If you had a bag with only 2 blue marbles, picking a red marble from the bag is

   certain   probable   impossible

2 If you had a bag with 2 black marbles and 9 pink marbles, picking a pink marble from the bag is

   certain   probable   impossible

Answer each question.

3 Kwan has a bag of toy fish. There are 8 red and 4 yellow fish. Without looking, is he more likely or less likely to pick a yellow fish? _____

4 What color do you predict you will spin more often? Why?

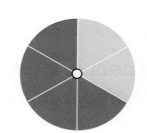

_____

_____

Solve.

5 Ashley uses the numbers 3, 7, and 9 to make different 2-digit numbers. What numbers can she make?

_____

Choose the best answer.

1 About how much does the glass hold?

I fluid ounce ⚪   8 fluid ounces ⚪   I quart ⚪   I gallon ⚪

2 What time is it?

11:15 ⚪   11:45 ⚪   12:15 ⚪   12:45 ⚪

3 How much money?

46¢ ⚪   55¢ ⚪   56¢ ⚪   66¢ ⚪

Test Prep

Solve.

4 30 people were on the bus. Some people got off at Smith Street. Now 14 people are on the bus. How many people got off at Smith Street?

_____ people

5 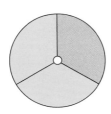 Which color will the spinner more likely land on? Explain how you know.

_____

Interpreting Data

The Giraffe Graph

by Sandra Liatsos

"My son," said the mother giraffe,

"very soon you'll grow bigger by half.

Each month we will measure

your height. What a pleasure

to show each new inch on a graph."

"I'll draw myself," said the giraffe,

"growing taller each month on my graph.

I'll soon be so tall

I'll go right off the wall,

and that will make both of us laugh."

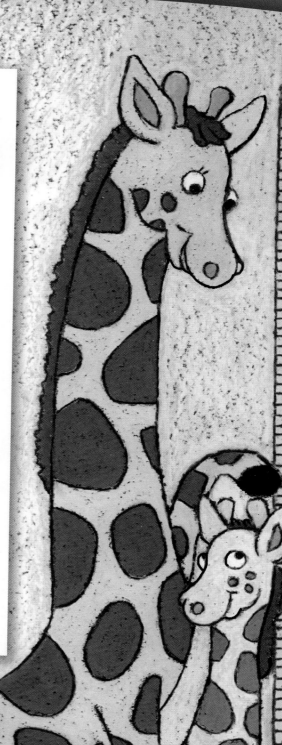

Math at Home

Dear Family,

I will read and interpret sets of data in Chapter 27. Here are my math words and an activity that we can do together.

Love, _____

My Math Words

range :

the difference between the greatest and least numbers

5 8 **4** 8 7 8 **9**

9 − 4 = 5

range = 5

mode :

the number that occurs most often

5 **8** 4 **8** 7 **8** 9

mode = 8

median :

the middle number when numbers are put in order from least to greatest

4 5 7 **8** 8 8 9

median = 8

Home Activity

Have your child think of a question to ask five people, such as "What is your favorite pet?" Then have your child record the results of the survey and discuss the information with you.

Favorite Pets	Tally	Total
🐱	I	1
🐶	III	3
🐦	I	1

Books to Read

Look for these books at your local library and use them to help your child learn about different ways to show data.

- **How Many Snails?** by Paul Giganti, Morrow, William, & Company, 1994.
- **X Marks the Spot** by Lucille Recht Penner, The Kane Press, 2002.
- **The Button Box** by Margarette S. Reid, Puffin Books, 1990.

www.mmhmath.com
For Real World Math Activities

© Macmillan/McGraw-Hill

Name_____

Learn You can describe a set of numbers by finding the range and mode of the numbers.

Math Words
range
mode

Al, Ken, Dennis, May, and Jan counted the letters in their names.

Al 2   Ken 3   Dennis 6   May 3   Jan 3

The range is the difference between the greatest and least numbers.

The mode is the number that you see most often.

Subtract to find the range.

$6 - 2 = 4.$

The range is __4__.

You see 3 most often.

The mode is __3__.

Try It Use these names. Answer each question.

 Kevin 5   Emma 4   Sara 4   Ty 2

1. How many letters are in the longest name? __5__

2. What is the range of the numbers? _____

3. Which number appears most often? _____

4. What is the mode of the numbers? _____

5. ✏ Write **About It!** How do you find the range for a set of numbers?

Practice Make towers with the following number of cubes. Find the median.

4. 4 cubes, 2 cubes, 9 cubes, 5 cubes, 7 cubes

> The median is the number that is in the middle.

Median: _____

5. 3 cubes, 1 cube, 8 cubes, 10 cubes, 6 cubes

Median: _____

6. 4 cubes, 2 cubes, 7 cubes, 6 cubes, 3 cubes

Median: _____

7. 5 cubes, 2 cubes, 13 cubes, 7 cubes, 10 cubes

Median: _____

8. 6 cubes, 1 cube, 6 cubes, 8 cubes, 8 cubes

Median: _____

Problem Solving — Use Data

Solve. Explain your answer.

9. Josh found bugs every day for 5 days. Here are the numbers for each day.

8   6   10   12   9

What is the median of the numbers?

_____

10. Amy collected stickers for 5 days. Here are the numbers for each day.

20   2   15   11   1

What is the median of the numbers?

_____

Math at Home: Your child practiced finding the median of a set of numbers.
Activity: Have your child arrange 8 books in 3 stacks. Ask him or her to record the numbers in each stack and then tell you what the median is.

Name_____

Coordinate Graphs

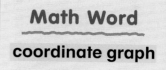

Learn A coordinate graph can show you where things are located.

This graph shows you where things are in a park. Find the sandbox.

- Always start at 0.
- First count to the right ⟶.
- Then count up ↑.
- To find the sandbox go to the right 1 and up 2.

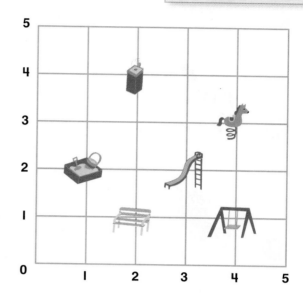

Try It Which thing would you find? Circle the answer.

	Right ⟶	Up ↑			
1	2	4	sandbox	(mailbox circled)	slide
2	4	3	spring horse	slide	swing set
3	2	1	bench	sandbox	slide
4	3	2	mailbox	slide	sandbox

5 **Write About It!** Where is the swing set? Tell how you would get there.

Practice
Where is each animal? Write the numbers.

	Right →	Up ↑
6 (giraffe)	3	3
7 (zebra)	____	____
8 (bird)	____	____
9 (elephant)	____	____
10 (monkey)	____	____

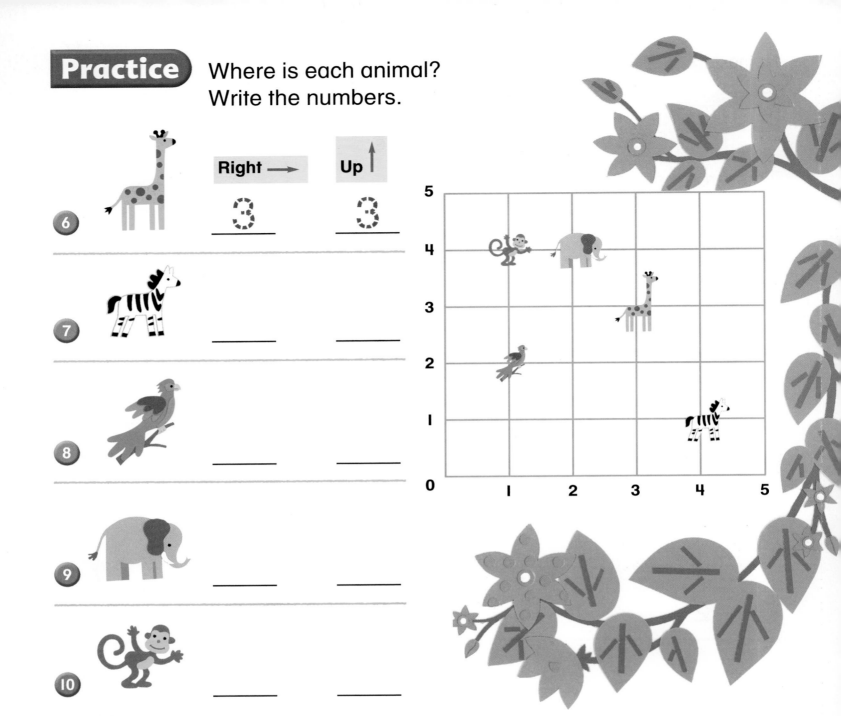

Problem Solving / Number Sense

11 An apple costs 75¢. Larry does not have enough money to buy it. He needs 10¢ more. How much money does Larry have? Explain.

_____

12 An orange costs 40¢. Sue needs 15¢ more. How much money does she have? Explain.

_____

Math at Home: Your child learned how to locate and write points on a grid.
Activity: Name an object on the grid on page 513. Have your child name the points where it is found using the words *go right* and *go up*.

514 five hundred fourteen

Learn You can use a line graph to compare data over time.

Adam wrote the daily high temperatures for four days.

Temperature	
Day	**°F**
Day 1	70°
Day 2	60°
Day 3	70°
Day 4	80°

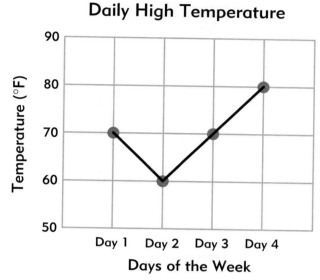

Daily High Temperature

Then Adam drew a dot for each day.
He connected the dots to make a line graph.

Try It Complete the line graph.

1. Anna measured her puppy every month for 4 months.

Puppy Height	
Month	**Height in Inches**
March	7
April	8
May	8
June	9

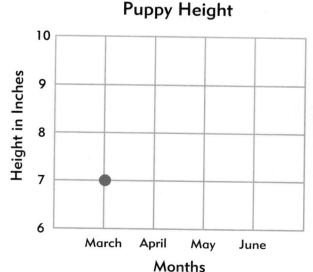

Puppy Height

2. ✏️ Write **About It!** Is the puppy bigger in June than in March? How do you know?

Practice Make a line graph to show the data.

> Draw a dot for each day.

Sam wrote the temperature at noon for four days.

Daily Noon Temperature	
Day	**Temperature**
Thursday	50°
Friday	55°
Saturday	60°
Sunday	50°

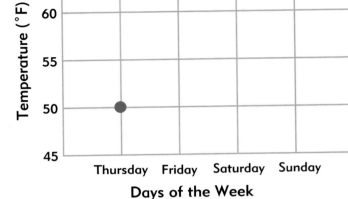

Daily Noon Temperature

Use the line graph above to solve.

3. Was it warmer at noon on Thursday or Friday?

_____

4. On which day was it the warmest at noon?

_____

Problem Solving — Use Data

 THINK SOLVE EXPLAIN

Use the Venn diagram to answer the questions.

5. How many children have a pet? _____

6. How many children have a sister and a pet? _____

7. How many children have a brother, sister, and a pet? _____

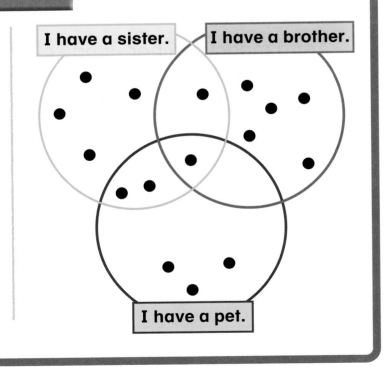

 Math at Home: Your child learned how to make and interpret line graphs.
Activity: Ask your child to explain how he or she made the graph on this page.

Name_____

Problem Solving Skill
Reading for Math

Boxes and Boxes

There are 5 trucks in the lot behind the supermarket. They bring fruits and vegetables from farms outside the city. The graph below shows what they brought.

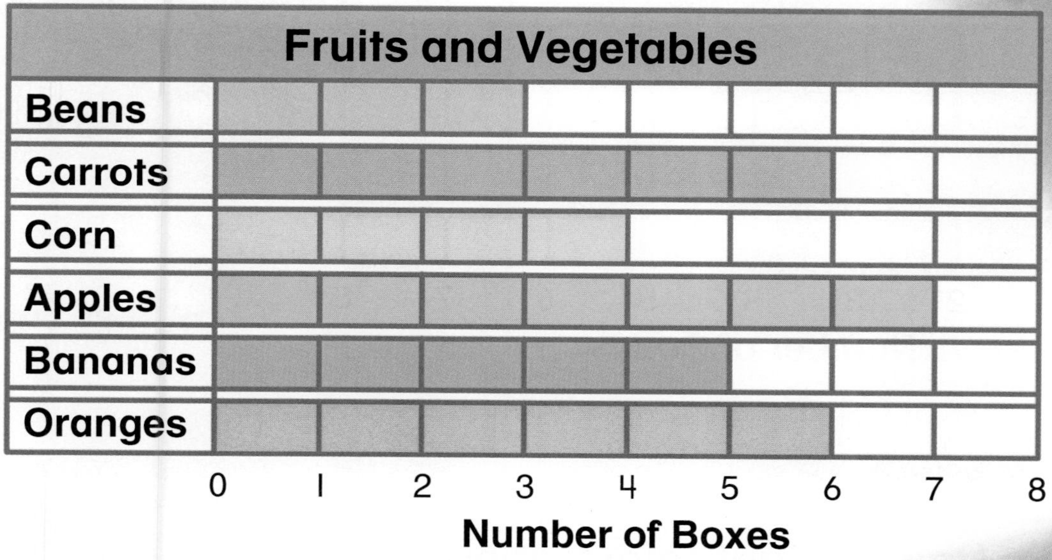

Fruits and Vegetables

Beans								
Carrots								
Corn								
Apples								
Bananas								
Oranges								

0   1   2   3   4   5   6   7   8

Number of Boxes

Problem Solving

Reading Skill **Important and Unimportant Information**

1. Was the number of trucks important to the story? _____

2. How many boxes of carrots were there? _____

3. How many boxes of bananas were there? _____

4. Are there more boxes of corn or apples? _____

© Macmillan/McGraw-Hill

More Boxes

Even though it's raining, more trucks arrived. The graph shows what they brought. Next week, the trucks will return to deliver more boxes.

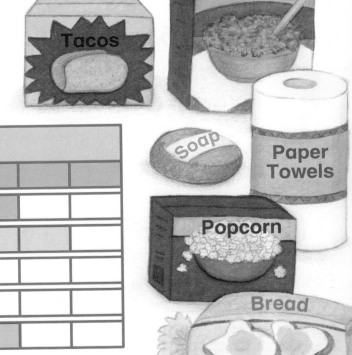

Problem Solving

Things to Buy

Bread									
Cereal									
Paper Towels									
Popcorn									
Soap									
Tacos									

0   1   2   3   4   5   6   7   8
Number of Boxes

 Reading Skill   **Important and Unimportant Information**

5 What information in the story was unimportant to help you read the graph?

_____

6 How many boxes of paper towels were there? _____

7 The trucks delivered 6 boxes of two items. What were they?

_____

Math at Home: Your child identified information that was not needed to answer questions.
Activity: Have your child cross out the unimportant sentences in each story.

Name_____

Problem Solving Practice

Solve.

1 How many children like to 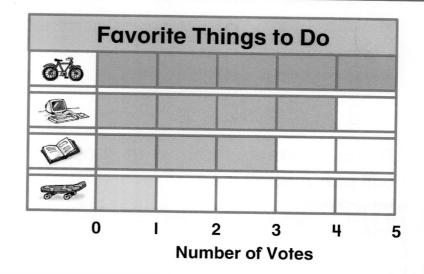 ?

_____ like to 🚲

Favorite Things to Do

🚲					
💻					
📖					
🛹					

0   1   2   3   4   5
Number of Votes

2 Do more children like The Frog Prince or Puss in Boots?

_____

How many more?

_____

Favorite Fairy Tales

Rapunzel							
Cinderella							
Puss in Boots							
The Frog Prince							

0   1   2   3   4   5   6   7
Number of Votes

THINK SOLVE EXPLAIN

Write a Story!

3 Write a question about this data. Answer your question.

_____

_____

_____

Favorite Pets

Dog						
Bird						
Cat						
Fish						

0   1   2   3   4   5   6
Number of Votes

Writing for Math

 Rosa puts these dots on a coordinate graph.

Right →	Up ↑
1	2
2	4
3	2

If she connects them what shape will she make?

Think

Always start at 0.

First count to the right →.

Then count up ↑.

Solve

Put the dots on the coordinate graph. Then connect the dots.

Rosa will make a _____.

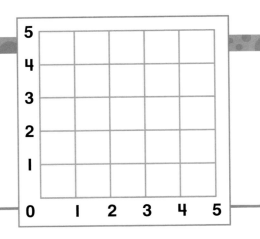

Explain

I can tell you how I found the points and drew the shape.

_____

_____

 e-Journal **www.mmhmath.com**
Write about math

Name_____

Use the data to answer each question.

1 What is the range of the numbers?

_____

2 What is the mode of the numbers?

_____

3 The dance was 2 hours long on Friday. How many tickets in all were sold on Friday and Monday?

_____

Dance Tickets Sold	
Monday	25
Tuesday	18
Wednesday	25
Thursday	19
Friday	14

Find each object. Complete to show where it is.

Right →	Up ↑

4 ● ___3___   ___4___

5 ⬡ _____ _____

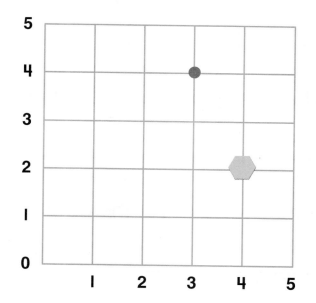

Assessment

© Macmillan/McGraw-Hill

Use this graph to answer problems 1–2.

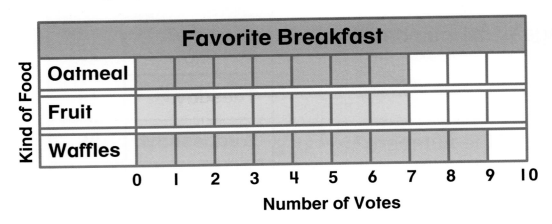

Favorite Breakfast

Kind of Food	
Oatmeal	
Fruit	
Waffles	

0 1 2 3 4 5 6 7 8 9 10

Number of Votes

1 How many like oatmeal?

7   9   10   11

○   ○   ○   ○

2 How many more like waffles than fruit?

1   2   7   9

○   ○   ○   ○

3 What is the range of these numbers?

2, 7, 6, 12, 5   _____

4 Write 3 facts about the graph.

_____

_____

_____

Favorite Color	
Green	★
Yellow	★ ★ ★ ★
Purple	★ ★
Red	★ ★ ★
Each ☆ stands for 2 votes.	

Exploring Multiplication and Division

Caterpillar Pete

by Sandra Liatsos

"If I could multiply times two,"

said Caterpillar Pete,

"I'd figure out

how many shoes

I needed for my feet.

On winter mornings every foot

could snuggle in its slipper.

On summer mornings in the sea

each foot could flip a flipper."

"If I could multiply times two,"

the caterpillar cried,

"I'd know how many shoes to buy

by counting just one side!"

Math at Home

Dear Family,

I will multiply and divide in Chapter 28. Here are my math words and an activity that we can do together.

Love, _____

My Math Words

multiply:

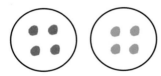

$2 \times 4 = 8$

divide:

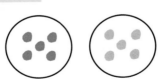

$10 \div 2 = 5$

Home Activity

Make a set of cards with addition sentences on them, such as $5 + 5 = 10$, and $2 + 2 + 2 + 2 = 8$.

Arrange 10 straws on the table in 2 groups of 5.

Ask your child to find the addition sentence that shows how many straws in all. Repeat the activity with other equal groups of straws.

© Macmillan/McGraw-Hill

Books to Read

In addition to these library books, look for the Time For Kids math story that your child will bring home at the end of this unit.

- **My Full Moon Is Square** by Elinor Pinczes, Houghton Mifflin, 2002.
- **The Doorbell Rang** by Pat Hutchins, Morrow, William & Company, 1992.
- **Time For Kids**

www.mmhmath.com
For Real World Math Activities

Name_____

**Explore
Equal Groups**

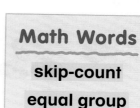
HANDS ON
Activity

Learn When there are equal groups,
you can skip-count to find the total.

Math Words

skip-count

equal group

_2_ , _4_ , _6_ , _8_ counters in all

You can make equal groups.
Put two counters in each group.
How many equal groups?

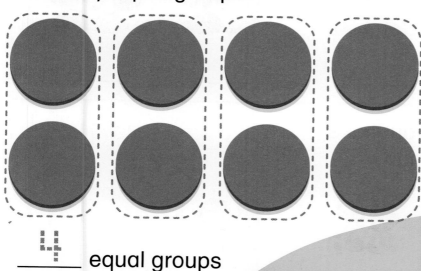

_4_ equal groups

Your Turn Use . Skip-count to find the total.

1 Make 2 groups of 4.

_8_ ● in all

2 Make 4 groups of 3.

_____ ● in all

3 Write **About It!** When can you skip-count to find how
many in all?

© Macmillan/McGraw-Hill

Chapter 28   Lesson 1

five hundred twenty-five **525**

Skip-count.
Write how many in all.

The bugs are in equal groups.

 4

3 _____   6 _____   9 _____   12 _____ in all

 5

_____   _____   _____   _____   _____ in all

 6

_____   _____   _____   _____   _____ in all

How many equal groups can you make?
Circle the groups. Write how many.

7

_____ equal groups

 8

_____ equal groups

 Math at Home: Your child learned to skip-count equal groups.
Activity: Give your child a pile of nickels and have him or her skip-count by 5s. Then ask the child to put the coins in piles to make equal groups.

526 five hundred twenty-six

Name_____

Repeated Addition and Multiplication

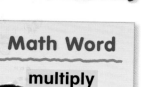
HANDS ON
Activity

Learn When groups are equal, you can use repeated addition to find how many in all. Or you can multiply .

Math Word

multiply

I can write an addition sentence.

__2__ + __2__ + __2__ + __2__ + __2__ = __10__

I can also write a multiplication sentence.

__5__ × __2__ = __10__

Five groups of 2 are 10.

Your Turn Use  to make equal groups. Add. Then multiply.

1 Make 2 groups of 4.

__4__ + __4__ = __8__

__2__ × __4__ = __8__

2 Make 3 groups of 2.

_____ + _____ + _____ = _____

_____ × _____ = _____

3 Make 4 groups of 5.

_____ + _____ + _____ + _____ = _____

_____ × _____ = _____

4  Write **About It!** How would you write a multiplication sentence for 3 + 3 + 3 + 3? Explain.

Write a multiplication sentence for each array.

Use an array to help you multiply.

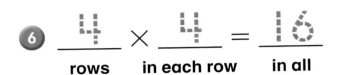

(6) $\underline{4} \times \underline{4} = \underline{16}$

rows   in each row   in all

(7) $\underline{} \times \underline{} = \underline{}$

rows   in each row   in all

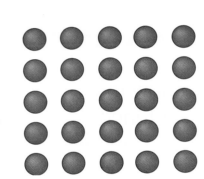

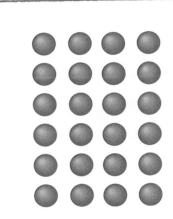

(8) $\underline{} \times \underline{} = \underline{}$

rows   in each row   in all

(9) $\underline{} \times \underline{} = \underline{}$

rows   in each row   in all

Problem Solving   Number Sense

Show Your Work

Draw a picture to solve.

THINK SOLVE EXPLAIN

(10) There are 2 rows of flowers. Each row has 4 flowers. How many flowers in all?

$\underline{} \times \underline{} = \underline{}$

rows   in each row   in all

Math at Home: Your child practiced writing multiplication sentences using multiplication arrays.
Activity: Draw 2 rows of 3 dots. Have your child write a multiplication sentence to show the array. ($2 \times 3 = 6$)

Name_____

Add or multiply.

$2 + 2 + 2 =$ 6

$3 \times 4 =$ ☐

$2 \times 5 =$ ☐

$8 \times 2 =$ ☐

$6 \times 4 =$ ☐

$3 \times 6 =$ ☐

$6 \times 2 =$ ☐

$5 + 5 + 5 =$ ☐

$2 + 2 + 2 + 2 =$ ☐

$5 \times 3 =$ ☐

$2 \times 4 =$ ☐

$5 \times 5 =$ ☐

$2 \times 2 =$ ☐

$1 + 1 + 1 + 1 =$ ☐

$4 \times 1 =$ ☐

Extra Practice

Circle the unit you would use to measure.

1 the amount of juice in a

(cups)   gallons

2 the length of a

inches   feet

3 how heavy a is

grams   kilograms

4 the amount of milk in a

ounces   gallons

5 the length of a

inches   yards

6 the weight of a

ounces   pounds

7 the amount of water in a

ounces   gallons

8 the length of a

inches   feet

9 the length of a

centimeters   meters

10 the height of a

inches   yards

11 Would you use inches or yards to measure the length of the playground? Explain why.

_____

 Math at Home: Your child practiced choosing the right unit to measure an object.
Activity: Choose an object in the kitchen. Ask your child to identify units for measuring its length and weight.

Name_____

Repeated Subtraction and Division

Learn You can use repeated subtraction to find the number of equal groups. Or you can divide.

Math Word
divide

Use . Separate 12 cubes into equal groups of 4 each. How many equal groups of 4 can you make?

Subtract 4 three times. Make 3 groups of 4. Write a division sentence.

$12 - 4 = 8$ $8 - 4 = 4$ $4 - 4 = 0$

You get ___3___ groups of 4.

$\underline{12} \div \underline{4} = \underline{3}$

Your Turn Use □. How many equal groups can you make? Subtract. Then divide.

1. Use 10 □.
   Subtract groups of 2.

   You get ___5___ groups of 2.

   $\underline{10} \div \underline{2} = \underline{5}$

2. Use 6 □.
   Subtract groups of 3.

   You get _____ groups of 3.

   $\underline{} \div \underline{} = \underline{}$

3. ✎ Write **About It!** How does subtracting equal groups help you divide?

Practice Divide. You can draw a picture to help.

You can divide to find equal shares.

5 12 flowers
6 equal groups
How many flowers in each group?

$$12 \div 6 = \underline{2}$$

6 16 bugs
4 equal groups
How many bugs in each group?

$$16 \div 4 = \underline{}$$

7 20 ants
4 equal groups of ants
How many ants in each group?

$$20 \div 4 = \underline{}$$

8 15 butterflies
3 equal groups
How many butterflies in each group?

$$15 \div 3 = \underline{}$$

✓ Spiral Review and Test Prep

9 Which number completes the multiplication?

$$5 \times 8 = \boxed{}$$

12   13   40   58
○   ○   ○   ○

10 What is the mode?

4, 3, 2, 3, 1

2   3   4
○   ○   ○

Math at Home: Your child divided to find equal shares.
Activity: Show 10 buttons. Have your child make 5 equal groups and tell how many are in each group. (2 buttons)

536 five hundred thirty-six

Problem Solving Strategy

Name_____

Use a Pattern • Algebra

You can use a pattern to help you solve problems.

A duck has 2 feet.
How many feet are on
6 ducks?

Problem Solving

Read

What do I already know?   _____ feet on 1 duck

What do I need to find out?   _____

Plan

I can find a pattern.

Solve

I can carry out my plan.
I can make a chart.

Number of Ducks	1	2	3	4	5	6
Number of Feet	2	4				

There are _____ feet on 6 ducks.

Look Back

What pattern do I see? _____

five hundred thirty-seven **537**

© Macmillan/McGraw-Hill

Use a number pattern to solve.

1 How many wings are on 5 butterflies?

There are _____ wings on 5 butterflies.
What pattern do I see?

Number of Butterflies	1	2	3	4	5
Number of Wings	4	8			

_____

2 How many fingers are on 5 gloves?

There are _____ fingers on 5 gloves.
What pattern do I see?

Number of Gloves	1	2	3	4	5
Number of Fingers	5				

_____

3 How many wheels are on 5 tricycles?

There are _____ wheels on 5 tricycles.
What pattern do I see?

Number of Tricycles	1	2	3	4	5
Number of Wheels	3				

_____

4 How many legs are on 6 crabs?

There are _____ legs on 6 crabs.
What pattern do I see?

Number of Crabs	1	2	3	4	5	6
Number of Legs	10					

_____

Math at Home: Your child solved problems by using number patterns.
Activity: Have your child continue one of the number patterns on this page.

Name_____

Butterfly Bingo

👥 2 players

You Will Need

2 🎲

15 🔴

15 🟤

▶ You and a partner take turns.
Choose blue or green for your counter.

▶ Toss both cubes.

▶ Multiply the two numbers and place your counter on the answer.

The first player to cover a row, column, or diagonal wins.

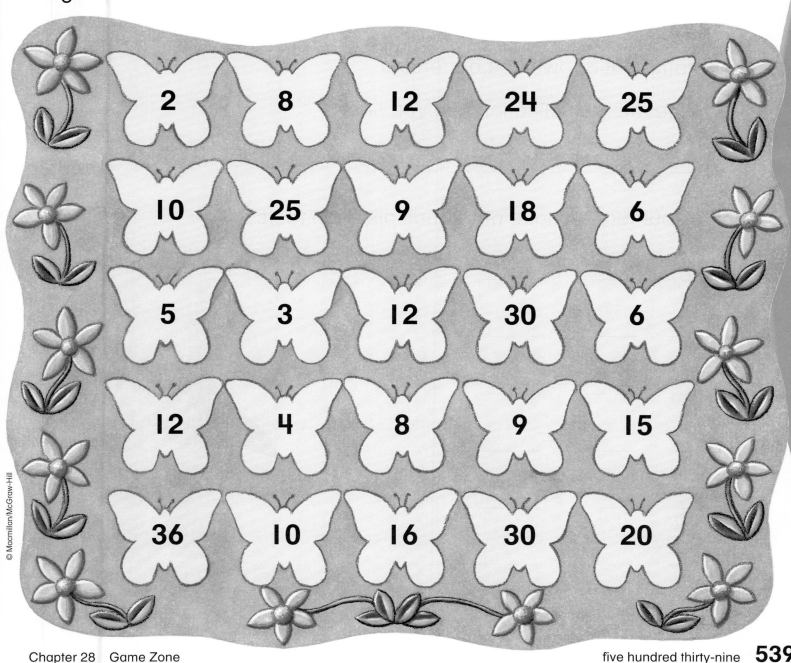

2	8	12	24	25
10	25	9	18	6
5	3	12	30	6
12	4	8	9	15
36	10	16	30	20

Technology Link

Model Multiplication • Computer

Use .

- Choose a mat to show one number.

- Stamp out 2 rows of 5 butterflies.

- What multiplication fact do you see?

$\underline{2} \times \underline{5} = \underline{10}$

- Stamp out 3 rows of 4 butterflies.

- Write the multiplication fact you see.

$\underline{3} \times \underline{4} = \underline{12}$

Use counters to make other multiplication facts.

_____ × _____ = _____

_____ × _____ = _____

_____ × _____ = _____

_____ × _____ = _____

_____ × _____ = _____

_____ × _____ = _____

For more practice use Math Traveler.™

Name_____

Write a multiplication sentence for each array.

1

2

_____ × _____ = _____   _____ × _____ = _____

Write how many groups.

3 16 beetles
    4 on each leaf

    _____ groups

    16 ÷ 4 = _____

Write how many in each group.

4 8 butterflies
    4 nets

    _____ in each group

    8 ÷ 4 = _____

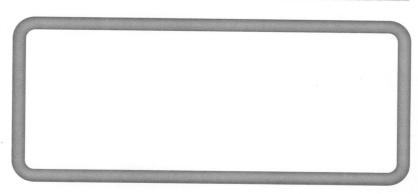

Use a pattern to solve.

5 There are 4 bugs. Each bug
    has 6 legs. How many legs in all?

    _____ legs

Number of Bugs	1	2	3	4
Number of Legs	6			

Assessment

© Macmillan/McGraw-Hill

Spiral Review and Test Prep
Chapters 1—28

Choose the best answer.

1 What is the mode of these numbers?

$$3, 10, 15, 4, 10, 9, 12$$

9	10	12	15
⚬	⚬	⚬	⚬

2 How many feet are there in a yard?

1	3	10	12
⚬	⚬	⚬	⚬

3 What is the value of the 5 in 450?

5	50	55	500
⚬	⚬	⚬	⚬

4 Add.

$$\begin{array}{r} 519 \\ +275 \\ \hline \end{array}$$

5 Multiply.

$$5 \times 2 = \underline{\hspace{1cm}}$$

6 If Mary takes a cube from the bag without looking, what cube is she more likely to pick?

_____ cube

Write to explain your answer.

D

Name _____

Two friends each eat one slice of the pie.

Write the fraction of the pie they eat.

Color the pie to show the fraction.

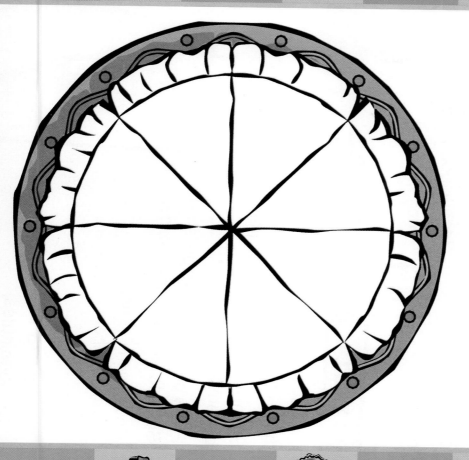

Fold down

TIME FOR KIDS

How Many Peaches?

Picking peaches with a friend is fun.

You can pick peaches to make a pie.

READ TOGETHER

Mom's Peach Pie Recipe

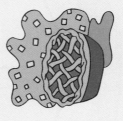

What You Will Need

A grown-up to help you

14 fresh peach halves
or canned peaches

$\frac{3}{4}$ cup sugar

$\frac{1}{4}$ cup flour

$\frac{1}{4}$ cup water or peach juice

2 tablespoons butter

2 tablespoons lemon juice

9-inch pie crust

What You Do

1. Mix sugar, butter, and flour to make crumbs.

2. Sprinkle half the mixture in the bottom of an unbaked pie crust.

3. Place peaches in the pie shell. Sprinkle with the lemon juice.

4. Cover with the rest of the crumbs.

5. Add fruit juice.

6. Bake at 375°F for 40 – 45 minutes.

Two friends picked 24 peaches in all.
6 peaches fit in a basket. They filled 4 baskets.

24 ÷ 6 = 4

Mom needs 12 peaches to make a pie.
She will use 2 baskets of peaches.

2 × 6 = 12

Name _____

Use Fractions to Make Decisions

Plan a pizza party for 6 children.
Each child may eat 2 or more slices.
3 children only eat mushroom pizza.
Each pizza has 8 slices.

1 Decide how many slices the children will eat.
Circle to show.

Show Your Work

2 How many slices of each kind of pizza did
you circle?

_____    _____

3 How many pizzas of each kind must you buy
for the party? Color to show.

_____

_____

© Macmillan/McGraw-Hill

Problem Solving

Plan a party for 12 children. Each child may have 2 or more party favors. There are 12 party favors to a box. Each box has one kind of party favor in it.

You Decide!

④ Decide how many party favors you will give the children. Color to show.

Show Your Work

Problem Solving

Your Decision!

⑤ How many of each kind of party favor will you need?

_____ 🪈   _____ 🪀

⑥ How many boxes of each must you buy for the party? Color to show.

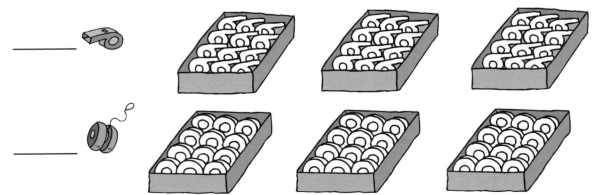

Math at Home: Your child applied fraction concepts to make decisions.
Activity: Show your child a group of two kinds of coins, such as 3 pennies and 3 nickels. Ask your child to write the fraction for the number of pennies.

Name_____

Math Words
Draw lines to match.

1 4 × 4 = 16

3 $\frac{1}{3}$

equal groups

multiplication sentence

fraction

Skills and Applications
Fractions (pages 473-484)

Examples

Fractions show equal parts of a whole.

$\frac{1}{4}$

1 of 4 equal parts

$\frac{2}{4}$

2 of 4 equal parts

4    $\overline{4}$

5    $\overline{2}$

Fractions can show equal parts of a group.

$\frac{3}{4}$ ← number of red fish
← total number of fish

6

$\overline{3}$

© Macmillan/McGraw-Hill

Skills and Applications

Multiplication and Division (pages 525–536)

Examples

You can use addition to help you multiply.

● ● ● ● ●
● ● ● ● ●

$5 + 5 = 10$

$2 \times 5 = 10$

⑦ $4 \times 2 = $ _____

⑧ $6 \times 3 = $ _____

⑨ $5 \times 5 = $ _____

⑩ $4 \times 3 = $ _____

How many in each group?

10 counters

5 groups

$10 \div 5 = 2$ in each group

⑪ 15 counters

3 in each group

$15 \div 3 = $ _____ groups

⑫ 16 counters

4 groups

$16 \div 4 = $ _____ groups

Problem Solving — Strategy

(pages 537–538)

THINK SOLVE EXPLAIN

⑬ Find the pattern.

Number of trucks	1	2	3	4	5
Number of wheels	4	8			

There are _____ wheels on 5 trucks.

What is the pattern?

_____

Math at Home: Your child learned about fractions, interpreting data, and multiplication and division.
Activity: Have your child use these pages to review fractions, interpreting data, and multiplication and division.

Name_____

Play a Game

Number of Bags   Number of Marbles

You Will Need

2
counters

Toss a cube on each mat to find
the number of bags and marbles.

Use the numbers to find
how many marbles in all.

Write a multiplication sentence.
You can use counters if you need to.

1 _____ × _____ = _____ marbles in all

2 _____ × _____ = _____ marbles in all

3 _____ × _____ = _____ marbles in all

4 _____ × _____ = _____ marbles in all

Portfolio   You may want to put this page in your portfolio.

Assessment

© Macmillan/McGraw-Hill

Remainder

Sometimes when you divide, you have a remainder.

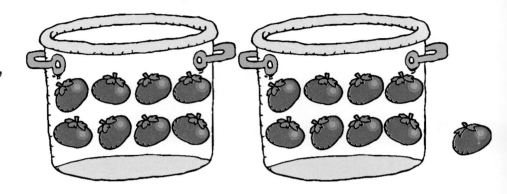

Show 17 tomatoes in 2 equal groups.

_8_ tomatoes in each group

_1_ tomato left $17 \div 2 = \underline{8} \ R \ \underline{1}$

Draw dots to show the equal groups and the remainder.

1 Show 13 tomatoes in 2 equal groups.

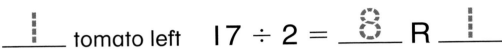

_____ tomatoes in each group

_____ tomato left $13 \div 2 = \underline{\qquad} \ R \ \underline{\qquad}$

2 Show 9 tomatoes in 4 equal groups.

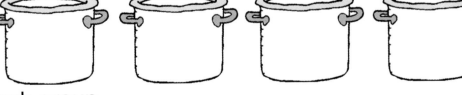

_____ tomatoes in each group

_____ tomato left $9 \div 4 = \underline{\qquad} \ R \ \underline{\qquad}$

3 Show 14 tomatoes in 3 equal groups.

_____ tomatoes in each group

_____ tomatoes left $14 \div 3 = \underline{\qquad} \ R \ \underline{\qquad}$

Picture Glossary

add (+) (page 7)

$$2 + 3 = 5$$

$$\begin{array}{r} 2 \\ +3 \\ \hline 5 \end{array}$$

area (page 381)

The area of this rectangle is 6 square units.

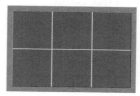

addend (page 17)

$$\begin{array}{r} 31 \\ +18 \\ \hline 49 \end{array}$$ ← addend
← addend

array (page 529)

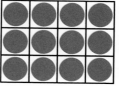

$$3 \times 4 = 12$$

after (page 97)

47 48 49

48 is after 47.

bar graph (page 189)

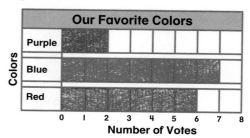

Our Favorite Colors

Purple, Blue, Red

Colors / Number of Votes

0 1 2 3 4 5 6 7 8

A.M. (page 171)

the hours from midnight to noon

It is 7:00 A.M.

before (page 97)

47 48 49

47 is before 48.

angle (page 355)

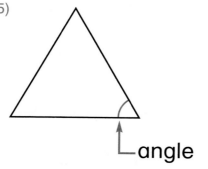

angle

between (page 97)

47 48 49

48 is between 47 and 49.

Glossary

© Macmillan/McGraw-Hill

Picture Glossary

calendar (page 177)

This is a calendar for September.

September						
S	M	T	W	T	F	S
				1	2	3
4	5	6	7	8	9	10
11	12	13	14	15	16	17
18	19	20	21	22	23	24
25	26	27	28	29	30	

chart (page 191)

Favorite Sports	
Soccer	7
Basketball	5
Football	4
Baseball	2

capacity (page 333)

the amount a container holds when filled

circle (page 355)

cent sign (¢) (page 115)

1¢   1 cent

compare (page 95)

5 is less than 7.   6 is equal to 6.   8 is greater than 4.

centimeter (cm) (page 323)

1 cm

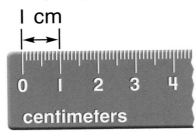

cone (page 353)

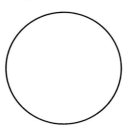

certain (page 493)

An event will definitely happen.

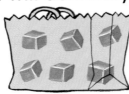

It is certain that you will pick a ▪.

congruent (page 373)

same size and same shape

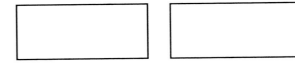

Picture Glossary

coordinate graph (page 513)

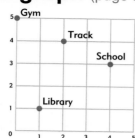

count back (page 33)

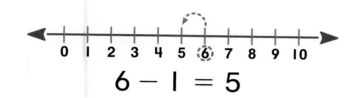

$$6 - 1 = 5$$

count on (page 19)

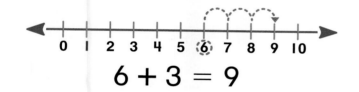

$$6 + 3 = 9$$

counting pattern (page 5)

2, 4, 6, 8, 10, 12, 14	Counting by 2s
3, 6, 9, 12, 15, 18, 21	Counting by 3s
5, 10, 15, 20, 25, 30, 35	Counting by 5s

cube (page 353)

cup (c) (page 335)

cylinder (page 353)

data (page 189)

information that is
collected for a survey

decimal point (page 133)

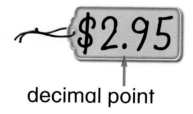

decimal point

degrees Celsius (°C) (page 343)

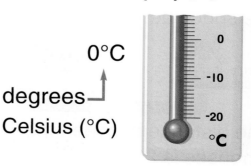

0°C

degrees
Celsius (°C)

Glossary

© Macmillan/McGraw-Hill

Picture Glossary

degrees Fahrenheit (°F) (page 343)

16°F

degrees ⌐ Fahrenheit (°F)

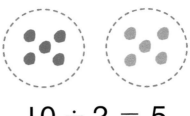

difference (page 7)

$17 - 9 = 8$

$$\begin{array}{r} 17 \\ -\ 9 \\ \hline 8 \end{array}$$

difference ⟶ 8

Subtract to find the difference.

digit (page 81)

any single figure used when representing a number

354

3, 5, and 4 are digits in 354.

dime (page 115)

10¢   10 cents

distance (page 323)

the space between two points

Jerry's house   Tina's house

200 yards

The distance between the 2 houses is 200 yards.

divide (page 533)

$10 \div 2 = 5$

dollar ($) (page 133)

$1.00   100 cents

doubles (page 53)

$7 + 7 = 14$

doubles plus 1 (page 53)

$7 + 8 = 15$

edge (page 353)

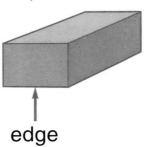

edge

Picture Glossary

equal groups (page 525)

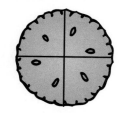

There are 4 equal groups of counters.

equal parts (page 473)

This pie is cut into 4 equal parts.

equal share (page 535)

Two children can share these crayons equally.

equally likely (page 495)

Without looking, it is equally likely that you will pick a ⬛ as a ⬛.

estimate (page 259)

$$47 + 22$$

$$50 + 20$$

about 70 ⟵ estimate

even (page 105)

2   4   6   8   10

expanded form (page 401)

another way of writing a number

364

$$300 + 60 + 4$$ ⟵ expanded form

face (page 353)

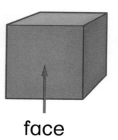

face

fact family (page 59)

$$5 + 6 = 11 \qquad 11 - 6 = 5$$
$$6 + 5 = 11 \qquad 11 - 5 = 6$$

flip (page 377)

a mirror image of a figure

© Macmillan/McGraw-Hill

Glossary

Picture Glossary

fluid ounce (fl oz) (page 335)

8 fluid ounces in a cup

half hour (page 155)

30 minutes

30 minutes = half hour

foot (ft) (page 321)

12 inches = 1 foot

hexagon (page 359)

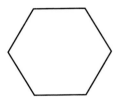

6 sides and 6 angles

fraction (page 473)

$\frac{1}{2}$ $\frac{1}{3}$ $\frac{1}{4}$ $\frac{3}{4}$

hour (page 155)

60 minutes

60 minutes = 1 hour

gallon (gal) (page 335)

4 quarts = 1 gallon

hour hand (page 155)

hour hand

half dollar (page 121)

50¢   50 cents

hundreds (page 397)

234

2 hundreds

Glossary

Picture Glossary

impossible (page 493)

an event that cannot happen

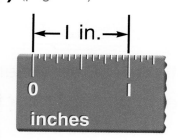

It is impossible to pick a ▪.

inch (in.) (page 319)

is equal to (=) (page 95)

35 is equal to 35.

35 = 35

is greater than (>) (page 95)

36 is greater than 35.

36 > 35

is less than (<) (page 95)

35 is less than 36.

35 < 36

key (page 195)

tells you what each symbol stands for

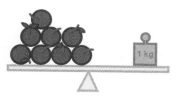

kilogram (kg) (page 341)

1 kilogram is about 8 apples.

length (page 319)

Length is how long something is.

less likely (page 495)

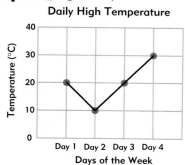

Without looking, it is less likely that you will pick a ▪.

line graph (page 515)

Glossary

Picture Glossary

line of symmetry (page 375)

a line on which a figure can be folded so that its two halves match exactly

median (page 511)

the middle number when numbers are put in order from least to greatest

6 7 8 9 10

The median is 8.

line plot (page 197)

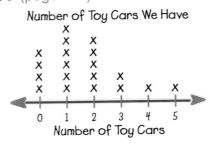

Number of Toy Cars We Have

meter (m) (page 323)

1 meter = 100 centimeters
1 meter is a little longer than a baseball bat.

liter (L) (page 339)

There are 1,000 milliliters in 1 liter.

milliliter (mL) (page 339)

This medicine dropper holds about 1 milliliter.

make a ten (page 55)

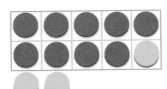

10 + 2

Make a ten to add 9 + 3.

minute (page 155)

1 minute

60 seconds = 1 minute

measure (page 317)

to find length, weight, capacity, or temperature

minute hand (page 155)

minute hand

Glossary

Picture Glossary

missing addend (page 41)

$$9 + \boxed{} = 14$$

The missing addend is 5.

nickel (page 115)

5¢   5 cents

mode (page 509)

the number that occurs most often in a set of data

4   7   10   36   7   2

The mode is 7.

number line (page 19)

0 1 2 3 4 5 6 7 8 9 10

month (page 177)

This calendar shows the month of September.

September						
S	M	T	W	T	F	S
				1	2	3
4	5	6	7	8	9	10
11	12	13	14	15	16	17
18	19	20	21	22	23	24
25	26	27	28	29	30	

number sentence (page 253)

$$6 + 8 = 14 \text{ or } 14 = 6 + 8$$

more likely (page 495)

Without looking, it is more likely that you will pick a ▪.

odd (page 105)

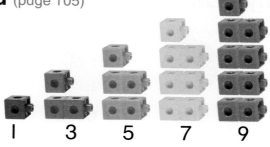

1   3   5   7   9

multiply (page 527)

$$2 \times 4 = 8$$

ones (page 77)

3 ones

© Macmillan/McGraw-Hill

Glossary

Picture Glossary

ordinal numbers (page 103)

numbers used to tell position

first   second   third

ounce (oz) (page 337)

One CD weighs about 1 ounce.

parallelogram (page 355)

4 sides and 4 angles

penny (page 115)

1 ¢   1 cent

pentagon (page 355)

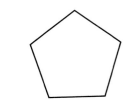

5 sides and 5 angles

perimeter (page 379)

the distance around a shape

pictograph (page 195)

Favorite Birds	
Crow	☺
Robin	☺ ☺ ☺ ☺
Jay	☺ ☺ ☺ ☺ ☺ ☺
Key: Each ☺ stands for 2 votes.	

picture graph (page 189)

Our Favorite Pets	
Dog	🐶🐶🐶
Turtle	🐢
Cat	🐱🐱🐱🐱

pint (pt) (page 335)

2 cups = 1 pint

place value (page 81)

the amount that each digit in a number stands for

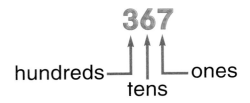

hundreds — tens — ones

Picture Glossary

P.M. (page 171)

the hours from
noon to midnight

It is 11:00 P.M.

pound (lb) (page 337)

The book weighs about
1 pound.

prediction (page 497)

a telling that something
will happen

probable (page 493)

an event that is more likely to happen

It is probable that you will pick a 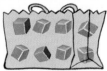.

pyramid (page 353)

quadrilateral (page 355)

4 sides and 4 angles

quart (qt) (page 335)

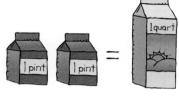

2 pints = 1 quart

quarter (page 121)

25¢   25 cents

quarter hour (page 159)

quarter hour = 15 minutes

range (page 509)

the difference between the least
number and the greatest number

4   7   10   36   7   2

greatest   least

36 − 2 = 34. The range is 34.

Glossary

Picture Glossary

reasonable (page 259)

A reasonable answer makes sense.

$$19 + 32 = 51$$
$$20 + 30 = 50$$

51 is a reasonable answer.

round (page 259)

to find the ten or hundred closest to a number

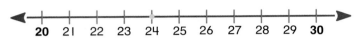

24 rounded to the nearest ten is 20.

rectangle (page 355)

4 sides and 4 angles

rule (page 21)

Add 3 is the rule.

Rule: Add 3

In	Out
10	13
20	23
30	33

rectangular prism (page 353)

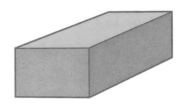

side (page 355)

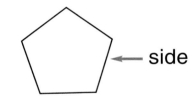

← side

regroup (page 209)

12 ones = 1 ten 2 ones

skip-count (page 5)

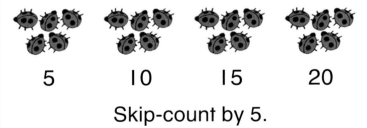

5   10   15   20

Skip-count by 5.

related facts (page 37)

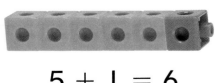

$$5 + 1 = 6$$
$$1 + 5 = 6$$

slide (page 377)

to move a figure horizontally, vertically, or diagonally

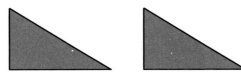

Picture Glossary

sphere (page 353)

square (page 355)

4 sides and 4 angles

subtract (−) (page 33)

5 − 3 = 2

sum (page 7)

sum
↓

3 + 2 = 5

The sum of 3 plus 2 is 5.

survey (page 191)

a way to collect data

Favorite Sports	
Soccer	卌 ||
Basketball	卌
Football	||||
Baseball	||

This survey shows favorite sports.

tally mark (page 191)

a mark used to record data

| = 1   卌 = 5

temperature (page 343)

a measure of hot or cold

The temperature is 79°F.

tens (page 77)

63

↑
6 tens

thousands (page 405)

1,253

↑
1 thousand

trapezoid (page 359)

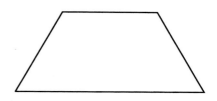

4 sides and 4 angles

Glossary

Picture Glossary

triangle (page 355)

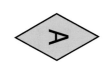

3 sides and 3 angles

week (page 177)

week →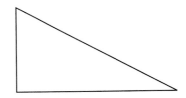

There are 7 days in a week.

turn (page 377)

a figure that is rotated around a point

yard (yd) (page 321)

3 feet = 1 yard

unit (page 363)

A   B   A   B   A   B   A   B

The things that repeat in
a pattern make a unit.

year (page 177)

one year

vertex (page 353)

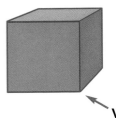

← vertex

Credits

Cover and title page photography: t. puppies-PhotoDisc; c. rug-Dot Box, Inc. for MMH.

Photography All photographs are by Macmillan/McGraw-Hill (MMH) except as noted below.

vi: Peter Brandt for MMH. vii: Peter Brandt for MMH. viii: Peter Brandt for MMH. ix: Mark E. Gibson/DRK Photo. xii: Peter Brandt for MMH. xvi: Peter Brandt for MMH. xvii: Peter Brandt for MMH. xviii: t. Image Source/Image State; b. Roger De La Harpe/Animals Animals; bkgd. PhotoDisc/Getty images. 3: t., b. EYEWire. 25: r. Kennan Ward/CORBIS. 26: t. PhotoDisc; c.t. Ingram Publishing; c.b., b. PhotoDisc. 37: Peter Brandt for MMH. 38: Peter Brandt for MMH. 45: Karl Weatherly/PhotoDisc. 55: Peter Brandt for MMH. 57: Peter Brandt for MMH. 64: t. Martial Colomb/PhotoDisc; c.t. Siede Preis/PhotoDisc; c.b. PhotoSpin; b. Artville. 68A: Georgette Douwma/Getty Images; inset Comstock. 68B: t.l. Georgette Douwma/Getty Images; t.r. Comstock; b.c. Fotosearch; b.r. Ryan McVay/Getty Images. 68C: t. Flip Nicklin/Minden Pictures; b.Flip Nicklin/Minden Pictures. 68D: t.l. Getty images; t.r. Getty Images; b. C Squared Studio/Getty Images; b.l.inset Georgette Douwma/Getty Images; b.r.inset Fotosearch. 77: Peter Brandt for MMH. 78: Peter Brandt for MMH. 79: Peter Brandt for MMH. 105: Peter Brandt for MMH. 108: t. Digital Vision/Getty Images; c.t. C Squared Studio/PhotoDisc. 115: t.r. Peter Brandt for MMH; c.r. Peter Brandt for MMH; b.l. Artville. 116: c.t. Ingram Publishing; c.b., b. Artville. 117: Peter Brandt for MMH. 119: t.l. Siede Preis/PhotoDisc; t.r. Peter Brandt for MMH; c.l. Artville; c.r. EYEWire. 123: t. Peter Brandt for MMH; c.b. Corel; b. PhotoDisc. 128: CORBIS. 129: t. Peter Brandt for MMH; c.r. PhotoDisc c.l. Siede Preis/PhotoDisc. 139: t.l., t.r. Peter Brandt for MMH; c., b. PhotoDisc. 140: PhotoDisc. 141: Peter Brandt for MMH. 142: t., b. Siede Preis/PhotoDisc; c.t. John A. Rizzo/PhotoDisc; c.b. Ingram Publishing. 145: b. PhotoDisc. 146A: Ariel Skelley. 146B: t. FotoSearch; b. Newscom. 146C: Joe Sohm/Chromosohm/Stock Connection/Picture Quest. 146D: t.l. Ariel Skelley; t.r. Newscom; b.l. FotoSearch; b.r.Joe Sohm/Chromosohm/Stock Connection/Picture Quest. 147: c. Craig Tuttle/Corbis; inset SuperStock. 148: t.r. Peter Brandt for MMH; c.l. John Shaw/Photo Researchers, Inc.; c.m.l. C Squared Studios/PhotoDisc; c.m.r. Jonelle Weaver/Getty Images; c.r. Kevin Schafer/CORBIS; b.l. John Shaw/Photo Researchers, Inc.; b.r. Jonelle Weaver/Getty Images. 155: Peter Brandt for MMH. 156: t. Peter Brandt for MMH; b.c. Stephen Simpson/Getty Images; b.r. Ariel Skelly/CORBIS. 166: Jeff Zaruba/Getty Images. 173: t. Peter Brandt for MMH, c. C Squared Studios/PhotoDisc; b. Photos.com. 174: t. C Squared Studios/PhotoDisc; c. EYEWire. 186: Siede Preis/PhotoDisc. 191: Peter Brandt for MMH. 192: Peter Brandt for MMH. 201: Jan Halaska/Photo Researchers, Inc. 202: Mark E. Gibson/DRK Photo. 209: Peter Brandt for MMH. 211: t. Peter Brandt for MMH. 213: Peter Brandt for MMH. 218: PhotoSpin. 222A: Peter Christopher/Masterfile; inset Jeff Greenberg/Index Stock. 222B: Ed Bock/CORBIS. 222C: t. Jeff Greenberg/Index Stock; b. Keith Brofsky/Getty Images. 222D: Mark Gibson/Index Stock. 223: c. Charles O'Rear/CORBIS; inset Astrid & Hanns Frieder-Michler/Photo Researchers, Inc. 263: Ann Purcell, Carl Purcell/Words & Pictures/PictureQuest. 264: t.c., b. PhotoDisc. 277: t. Peter Brandt for MMH. 278: Peter Brandt for MMH. 303: Tomas del Amo/Index Stock Imagery. 304: t., b. PhotoDisc. 308A: Keren Su/Getty Images; inset PhotoDisc. 308B: CORBIS. 308C: t. Scott McKinley/Getty Images; b. Daryl Balfour/Getty Images. 308D: t.l. Kennan Ward/CORBIS; t.r. Tom Brakefield/CORBIS; b. Chris Everard/Getty Images. 317: t.l. PhotoDisc; t.r. Peter Brandt for MMH; c.b., b. Artville. 318: t., t.c., b.c. Artville; c. David Toase/PhotoDisc. 319: t.r. Peter Brandt for MMH; c. Artville; b. Scott Harvey for MMH. 321: t. Peter Brandt for MMH. 322: Peter Brandt for MMH. 323: t.r. Peter Brandt for MMH; row 1, 3 Artville; row 2 StockByte; row 4 PhotoDisc. 324: t.r. Peter Brandt for MMH; c.t. Geostock/PhotoDisc; t.c.r. Siede Preis/PhotoDisc; l.c. C Squared Studios/PhotoDisc; b.c.r. PhotoDisc; b.r. Corel. 333: t. Peter Brandt for MMH; b.r. C Squared Studios/PhotoDisc. 334: t. Peter Brandt for MMH; b.r. Ryan McVay/PhotoDisc. 335: Peter Brandt for MMH. 336: t.l. PhotoDisc; b.l. Burke/Triolo/Brand X Pictures/PictureQuest; b.r. John Campos/FoodPix. 337: t.l. PhotoSpin; t.r. Artville; b.r. Dot Box, Inc. for MMH. 338: t.l. PhotoDisc; t.r. Peter Brandt for MMH; c.r. EYEWire; b.l. Index Stock Imagery;. 339: t. Peter Brandt for MMH; c.r. PhotoSpin. 340: t. Peter Brandt for MMH; t.l., b.l.: PhotoDisc; b.r. Index Stock Imagery. 341: t.l. Michael Matisse/PhotoDisc; b.l. Scott Harvey for MMH; b.r. EYEWire. 342: t.r. Peter Brandt for MMH; t.l. Dot Box, Inc. for MMH; t.c.l. b.c.l. PhotoDisc. 349: t.c. PhotoDisc; t.r. EYEWire; b.c., b.r. Ingram Publishing. 357: Peter Brandt for MMH. 359: Peter Brandt for MMH. 363: Peter Brandt for MMH. 373: Peter Brandt for MMH. 374: Peter Brandt for MMH. 375: Peter Brandt for MMH. 376: Peter Brandt for MMH. 377: Peter Brandt for MMH. 379: Peter Brandt for MMH. 381: Peter Brandt for MMH. 388A: Cover National Geographic; inset Siede Preis/PhotoDisc. 388B: b.l. Bettmann/CORBIS. 388B-C bkgd. Grant Heilman/Index Stock. 388C: r. Courtesy Jefferson National Parks Association. 388D: AP. 389: l. Michael Black/Bruce Coleman; l. inset Phil Degginger/Bruce Coleman. Inc.; r. M.Timothy O'Keefe/Bruce Coleman, Inc.; r. inset Deborah Davis/Photo Edit. 390: Peter Brandt for MMH. 397: Peter Brandt for MMH. 438: Peter Brandt for MMH. 440: Peter Brandt for MMH. 454: Peter Brandt for MMH. 456: Peter Brandt for MMH. 460: t. Stockbyte/PictureQuest; c.t. Siede Preis/PhotoDisc; c.b. Photos.com. 464A: cvr. SuperStock; inset PhotoDisc. 464B: Creatas. 464C: t. Brand X Pictures; b.l. Superstock; b.r. Comstock. 464D: t.l. Ablestock; t.c., t.r. PhotoDisc; b.l. C Squared Studio/Getty Images; c.l. Photospin; c.r. PhotoDisc; r.c.t., r.c.b., b.r. Ablestock. 465: c. Nancy Sefton/Photo Researchers, Inc.; inset Ken Karp for MMH. 466: Miep Van Damm/Masterfile. 479: Peter Brandt for MMH. 480: Peter Brandt for MMH. 495: Peter Brandt for MMH. 496: Peter Brandt for MMH. 497: Peter Brandt for MMH. 498: Peter Brandt for MMH. 506: PhotoDisc. 512: Peter Brandt for MMH. 520: Peter Brandt for MMH. 525: Peter Brandt for MMH. 526: Peter Brandt for MMH. 527: Peter Brandt for MMH. 528: Peter Brandt for MMH. 529: Peter Brandt for MMH. 533: Peter Brandt for MMH. 535: Peter Brandt for MMH. 536: Peter Brandt for MMH. 542A: cvr. G. Bliss/Masterfile; inset Bob Kris/CORBIS. 542B: t. Bob Kris/CORBIS; b. PhotoDisc. 542C: Rubberball. Martin B. Withers/CORBIS; hickory leaf (t.r.) Hal Horwitz/Getty Images.

Illustration Bernard Adnet: 473, 474. Farah Aria: 233, 234, 235, 236, 269, 431. Martha Aviles: 169. Kristen Barr: 195, 196, 523. Jennifer Beck-Harris: 58, 275, 276, 413, 415, 416, 451, 452. Shirley Beckes: 9, 10, 11, 14, 35, 36, 159, 173, 281, 282, 346, 421, 422, 475, 476, 481, 482, 487. Sarah Beise: 135, 137, 138, 403, 404, 469, 470. Carly Castillon: 25, 26, 117, 118, 152, 199, 207, 227, 239, 240, 241, 242, 297, 298, 361, 362, 401, 402, 435, 436, 493, 494, 500. Randy Cecil: 271, 272, 279, 280, 287, 293, 294, 447. Chi Chung: 21, 22, 200, 201. Nancy Coffelt: 5, 6, 63, 64, 299, 300, 477, 478, 507, 515. Diana Craft: 65, 66, 183, 265, 273, 274, 385. Lynn Cravath: 39, 40, 163, 164, 171, 172, 441, 442, 443. Mike Dammer: 2, 16, 32, 45, 52, 68, 76, 94, 114, 127, 132, 170, 174, 185, 188, 192, 203, 206, 208, 222, 230, 247, 252, 266, 292, 304, 314, 316, 324, 332, 352, 364, 367, 372, 376, 378, 384, 391, 396, 409, 410, 412, 414, 432, 448, 472, 495, 498, 506, 508, 524, 542, 543, 544, 545, 548, g4, g5, g8, g10, g11, g14. Nancy Davis: 193, 194, 204. Sarah Dillard: 81, 92, 103, 104. Kathi Ember: 95, 96, 165, 214, 215, 216, 253, 254, 405, 406, 489, 490. Buket Erdogan: 351. Dagmar Fehlau: 61, 62, 161, 162, 177, 178, 295, 296, 353. Brian Fujimoro: 179, 180, 251. Barry Gott: 41, 42, 85, 86, 89, 90, 99, 100, 191, 192, 305. Steve Haskamp: 83, 301, 302, 313, 314. Eileen Hine: 143. Tim Huhn: 33, 34, 97, 98, 259, 260, 417, 418, 423, 424, 457, 458. Melissa Iwai: 51, 197, 198, 231, 232. Jong Un Kim: 93. Richard Kolding: 87, 88. Erika LeBarre: 331, 538, 539. Chris Lench: 133, 134. Rosanne Litzinger: 261, 262. Lori Lohstoeter: 461, 462. Margeau Lucas: 15, 113. Lyn Martin: 53, 54, 109, 112, 483, 484, 531, 532. Christine Mau: 131. Debra Melmon: 157, 158. Laura Merer: 27, 517, 518, 547. Pat Meyers: 371. Edward Miller: 309, 310, 335, 337, 338, 341, 347, 368. Taia Morley: 343, 383, 519, 526, 528, 534. Keiko Motoyama: 175, 380. Christina Ong: 187. Laura Ovresat: 69, 70, 107, 108, 125, 127, 255, 256. Liz Pichon: sponsor critters, 31, 101, 102, 325, 326, 327, 328. Jen Rarey: 379. Mick Reid: 17, 18, 59, 60, 257, 258, 399, 400, 419, 501, 503. Stephanie Roth: 85, 86. Christine Schneider: 283, 284, 291, 491. David Sheldon: 73, 217, 245, 246, 313, 407, 408, 425, 433, 434, 444, 485, 486. Janet Skiles: 23, 24, 509, 510. Marsha Slomowitz: 119, 120, 243, 244. Ken Spengler: 219, 220, 499. Maribel Suarez: 355, 356, 449, 450. Susan Synarski: 7, 8, 12, 48, 189, 190, 285, 286, 459. Mary Thelen: 513, 514. Pamela Thomson: 3, 4, 43, 44, 181, 182, 365, 366, 427.

Acknowledgments

The publisher gratefully acknowledges permission to reprint the following copyrighted material:

Bushy-Tailed Mathematicians from COUNTING CATERPILLARS AND OTHER MATH POEMS by Betsy Franco. Copyright © 1998 by Betsy Franco. Published by Scholastic, Inc. Reprinted by permission.

Caterpillar Pete from POEMS TO COUNT ON by Sandra Liatsos. Copyright © 1995 by Sandra O. Liatsos. Published by Scholastic, Inc. Reprinted by permission.

Going To Bed. http://www.headstart.lane.or.us/education/activities/music/songs-fingerplays.html. Reprinted by permission.

Money Rhymes. http://www.canteach.ca/elementary/songspoems70.html. Reprinted by permission.

Skip-Count Song. http://www.canteach.ca/elementary/songspoems72.html. Reprinted by permission.

The Giraffe Graph from POEMS TO COUNT ON by Sandra Liatsos. Copyright © 1995 by Sandra O. Liatsos. Published by Scholastic, Inc. Reprinted by permission.

Twelve Little Rabbits. http://www.headstart.lane.or.us/education/activities/music/songs-fingerplays.html. Reprinted by permission.

The book covers listed below are reprinted with the permission of the following publishers:

Charlesbridge: THE COIN COUNTING BOOK

HarperCollins Publishers: GAME TIME!; IF YOU GIVE A PIG A PANCAKE; JUMP, KANGAROO, JUMP!; MISSING MITTENS; 100 SCHOOL DAYS; ONE LIGHTHOUSE, ONE MOON; ROOM FOR RIPLEY

HarperCollins UK: I SPY TWO EYES

Houghton Mifflin: SO MANY CATS!

Penguin Putnam Publishers: THE BUTTON BOX; HANNAH'S COLLECTIONS; TWENTY IS TOO MANY

Random House Publishers: HOW BIG IS A FOOT?

Scholastic, Inc. Publishers: THE CASE OF THE SHRUNKEN ALLOWANCE; THE GRAPES OF MATH

Book Cover for HOW MANY TEETH? reprinted by permission of the estate of Paul Galdone.

Book Cover for MOIRA'S BIRTHDAY by R. Munsch and M. Martchenko. Copyright © Michael Martchenko, artwork 1987. Reprinted with permission of Annick Press.

Book Cover for MRS. McTATS AND HER HOUSEFUL OF CATS reprinted with the permission of Margaret K. McElderry Books, an imprint of Simon & Schuster Children's Publishing Division, by Alyssa Satin Capucilli, illustrated by Joan Rankin. Illustrations copyright © 2001 Joan Rankin.

Book Cover for ROUND AND SQUARE by Miriam Schlein and Linda Bronson from Mondo's BOOKSHOP Literacy Program. Text copyright © 1999, 1952 by Miriam Schlein. Illustrations copyright © 1999 by Linda Bronson, reprinted by permission of Mondo Publishing, 980 Avenue of the Americas, New York, New York, 10018. All rights reserved.

Book Cover for WHAT'S NEXT, NINA? by Sue Kassirer, illustrated by Page Eastburn O'Rourke. Copyright © 2001 The Kane Press.

ACKNOWLEDGMENTS

The publisher gratefully acknowledges permission to reprint the following copyrighted material:

Bushy-Tailed Mathematicians from COUNTING CATERPILLARS AND OTHER MATH POEMS by Betsy Franco. Copyright © 1998 by Betsy Franco. Published by Scholastic, Inc. Reprinted by permission.

Caterpillar Pete from POEMS TO COUNT ON by Sandra Liatsos. Copyright © 1995 by Sandra O. Liatsos. Published by Scholastic, Inc. Reprinted by permission.

Going To Bed. http://www.headstart.lane.or.us/education/activities/music/songs-fingerplays.html. Reprinted by permission.

Money Rhymes. http://www.canteach.ca/elementary/songspoems70.html. Reprinted by permission.

Skip-Count Song. http://www.canteach.ca/elementary/songspoems72.html. Reprinted by permission.

The Giraffe Graph from POEMS TO COUNT ON by Sandra Liatsos. Copyright © 1995 by Sandra O. Liatsos. Published by Scholastic, Inc. Reprinted by permission.

Twelve Little Rabbits. http://www.headstart.lane.or.us/education/activities/music/songs-fingerplays.html. Reprinted by permission.

The book covers listed below are reprinted with the permission of the following publishers:

Charlesbridge: THE COIN COUNTING BOOK

HarperCollins Publishers: GAME TIME!; IF YOU GIVE A PIG A PANCAKE; JUMP, KANGAROO, JUMP!; MISSING MITTENS; 100 SCHOOL DAYS; ONE LIGHTHOUSE, ONE MOON; ROOM FOR RIPLEY

HarperCollins UK: I SPY TWO EYES

Houghton Mifflin: SO MANY CATS!

Penguin Putnam Publishers: THE BUTTON BOX; HANNAH'S COLLECTIONS; TWENTY IS TOO MANY

Random House Publishers: HOW BIG IS A FOOT?

Scholastic, Inc. Publishers: THE CASE OF THE SHRUNKEN ALLOWANCE; THE GRAPES OF MATH

Book Cover for HOW MANY TEETH? reprinted by permission of the estate of Paul Galdone.

Book Cover for MOIRA'S BIRTHDAY by R. Munsch and M. Martchenko. Copyright © Michael Martchenko, artwork 1987. Reprinted with permission of Annick Press.

Book Cover for MRS. McTATS AND HER HOUSEFUL OF CATS reprinted with the permission of Margaret K. McElderry Books, an imprint of Simon & Schuster Children's Publishing Division, by Alyssa Satin Capucilli, illustrated by Joan Rankin. Illustrations copyright © 2001 Joan Rankin.

Book Cover for ROUND AND SQUARE by Miriam Schlein and Linda Bronson from Mondo's BOOKSHOP Literacy Program. Text copyright © 1999, 1952 by Miriam Schlein. Illustrations copyright © 1999 by Linda Bronson, reprinted by permission of Mondo Publishing, 980 Avenue of the Americas, New York, New York, 10018. All rights reserved.

Book Cover for WHAT'S NEXT, NINA? by Sue Kassirer, illustrated by Page Eastburn O'Rourke. Copyright © 2001 The Kane Press.